THE ARRL Tec[illegible] LI[illegible] MANUAL

For Novice Class Licensees

■ **Edited By**

Larry D. Wolfgang, WR1B
Joel P. Kleinman, N1BKE
Jim Kearman, KR1S

■ **Production Staff**

Deborah Strzeszkowski
Joe Shea
David Pingree, N1NAS
Steffie Nelson, KA1IFB,
Jodi Morin
Sue Fagan
Michelle Bloom, WB1ENT

■ **Published by:**
The American Radio Relay League
Newington, CT 06111-1494

Printed in USA

ISBN: 0-87259-418-1

Second Edition
First Printing

Notes to Those Preparing for the Technician Exam

You can obtain a Technician class Amateur Radio license with full privileges on all the amateur VHF and UHF bands without taking a Morse code exam. Chapter 1 contains more information.

If you already have a Novice license, or are taking a Novice course to prepare for that exam, this book and *The FCC Rule Book*, also published by ARRL, are all you need to prepare for the Technician exam. **If you do not have a Novice license**, we suggest you use *Now You're Talking! All you Need to Get Your First Ham Radio License* instead of this book. *Now You're Talking!* is a complete course in Novice *and* Technician theory, regulations and operating practice.

This book may be used for Technician (Element 3A) written exams given beginning July 1, 1993. The book may be used for exams given until July 1, 1997. (This ending date assumes no FCC rules changes to the licensing structure or privileges for the Technician license, which would force the Volunteer Examiner Coordinator's Question Pool Committee to modify the question pool printed in Chapter 10 of this book.)

Foreword

Congratulations! You have made the decision to upgrade from a Novice Amateur Radio license to a Technician license. You are about to take another step into the exciting world of Amateur Radio.

The Technician license will give you a wide range of new operating privileges. With a Technician license, you can roam all the spectrum allocated to the Amateur Service above 30 MHz — including the very popular 2-meter band.

League staff has worked diligently to gather the material you must master to pass the written examination for a Technician class license. Every question that you might be asked is included in this book. The text explains every technical concept in clear, easy to understand terms. The material is presented in a format that will help you learn in a classroom setting or by studying on your own.

In addition to *The ARRL Technician Class License Manual for Novice Class Licensees*, you'll want to have a copy of *The FCC Rule Book*, which is also published by the ARRL. *The FCC Rule Book* covers FCC regulations, with appropriate explanations.

This second edition of *The ARRL Technician Class License Manual* is not just the product of ARRL staff work. Readers of the first edition sent comments and suggestions. You can do your part to help with later printings and editions of this book, too. How? First use this book to prepare for your exam. After you have studied the material, write your suggestions (or any corrections you think need to be made) on the Feedback Form at the back of this book. Then send the form to us. Your input is important to help us improve our study materials.

David Sumner, K1ZZ
Executive Vice President, ARRL
Newington, Connecticut

HOW TO USE THIS BOOK

To earn a Technician class Amateur Radio License, you will want to know some basic electronics theory, and the rules and regulations governing the Amateur Radio Service, as contained in Part 97 of the FCC Rules. You do not have to know the international Morse code to pass the Technician exam. If you pass the 5-WPM code test, however, you will be granted additional operating privileges. Chapter 1 has more information.

This book provides a brief description of the Amateur Radio Service, and the Technician class license in particular. The major portion of the book is designed to guide you, step by step, through the theory you must know to pass the Element 3A license exam. If you hold a Novice class license, you must pass Element 3A to earn the Technician license. The material is presented in a manner that closely follows the study guide, or syllabus, printed at the end of Chapter 1. Chapter 10 contains the complete Element 3A Question Pool, used on Technician exams. It also includes the multiple-choice answers and distractors that will be used on these exams.

At the beginning of each chapter, you will find a list of **Key Words** that appear in that chapter, along with a simple definition for each word or phrase. As you read the text, you will find these words printed in **boldface type** the first time they appear. You may want to refer to the **Key Words** list at the beginning of the chapter when you come to a **boldface** word, to learn the definition. At the end of the chapter you may also want to go back and review those definitions. Appendix C is a glossary of all the key words used in the book. They are arranged in alphabetical order for your convenience as you review before the exam.

The Question Pool

As you study the material you will be instructed to turn to sections of the question pool in Chapter 10. Be sure to use these questions to review your understanding of the material at the suggested times. This will break the material into bite-sized pieces, and make it easier for you to learn. Do not try to memorize all of the questions. That will be impossible! Instead, by using the questions for review, you will be familiar with them when you take the test, but you will also understand the electronics theory behind the questions.

Most people learn more when they are actively involved in the learning process. Turning to the questions and answers helps you be actively involved, and will help you learn faster. If you would like the correct answers marked in the question pool, you can mark them as you study. This will reinforce the material in your mind. Make an asterisk or check mark in the left margin next to the correct answer. Then you can cover those marks with a slip of paper for review later. Paper clips make excellent place markers to help you find your spot in the text, question pool and answer key.

The correct answers to the questions are given in the answer key printed at the end of Chapter 10. Also printed with the answer key are page references indicating where you can turn to the text for a quick review of the material that explains each question.

Other ARRL Study Materials

In addition to this book, you will want to purchase a copy of the ARRL publication *The FCC Rule Book*, which covers all the rules and regulations you'll need to know. The ARRL also has *Now You're Talking!*, designed for beginners, which covers the theory and rules needed to pass both the Element 2 and Element 3A written tests.

If you choose to learn the Morse code to earn Novice high-frequency (HF) band privileges, the ARRL has materials to help you. *Your Introduction to Morse Code* is a companion cassette-tape course that teaches the Morse code letter by letter and provides practice at 5 WPM. ARRL also offers a series of cassette tapes, *Increasing Your Code Speed,* to help you go from 5 to 10 WPM, 10 to 15 WPM and 15 to 22 WPM. There are also two tapes for practice with 13 to 14 WPM code using standard character timing.

Even with the tapes, you'll want to tune in to the code-practice sessions transmitted by W1AW, the ARRL Headquarters station. The W1AW code-practice sessions are transmitted to assist you in learning the code. For more information about W1AW or how to order any ARRL publication or code tape, write to: ARRL, 225 Main St, Newington, CT 06111-1494.

WHEN TO USE THIS BOOK

Amateur-license-exam questions are written by a Question Pool Committee (QPC). The QPC consists of representatives from the various Volunteer Examiner Coordinators (VECs).

The QPC has established the following schedule for new question pools:

Pool	*Effective Date*
Novice (Element 2)	July 1, 1997*
Technician (Element 3A)	July 1, 1997*
General (Element 3B)	July 1, 1994
Advanced (Element 4A)	July 1, 1995*
Extra (Element 4B)	July 1, 1996*

*Tentative date

As the schedules for the question pools are confirmed, the information will be published in *QST*. New editions of ARRL License Manuals that cover the new question pools will be available before the effective dates of the new pools.

CODE EXAMINATION CREDIT FOR HANDICAPPED PERSONS

Effective February 14, 1991, FCC adopted new rules pertaining to Morse code examinations for handicapped persons. Applicants must have a current FCC amateur license that required a code test. If the license has expired, it must be within the grace period for renewal. Valid licenses include:

- Novice
- Technician plus a CSCE for passing Element 1A or 1B
- Technician issued before February 14, 1991
- General
- Advanced

Effective August 1, 1992, applicants must use a revised Form 610. The revised form has a date of march 1992 or later printed at the bottom of each page. The "PHYSICIAN'S CERTIFICATION OF DISABILITY" section of the new form must be completed and signed by a person licensed to practice in a place where the amateur service is regulated by the FCC as a doctor of medicine (MD) or doctor of osteopathy (DO). After the physician signs the Form 610, the applicant must also sign in the same section, granting the physician permission to disclose medical information relevant to the application. To obtain a new Form 610, send a business-sized, self-addressed, stamped envelope to: Federal Communications Commission, Forms Distribution Center, 2803 52nd Ave, Hyattsville, MD 20781 (specify "Form 610" on the envelope), or: Form 610, ARRL, 225 Main St, Newington, CT 06111-1494.

TABLE OF CONTENTS

1 The Technician Class License

Many people regard the Technician license as a good starting point in Amateur Radio. After all, with a Technician license comes the privilege of operating through repeaters on the popular 2-meter band. In 1991, FCC removed the Morse code requirement for the Technician license.

If you have put off earning your Technician license because you thought the electronics theory was too difficult, put those thoughts aside right now! Once you make the commitment to study and learn what it takes to pass the exam, you *will* be able to do it. It often takes more than one try to pass the exam, but many amateurs do pass on their first try. The key is that you must make the commitment, and be willing to study. With this book, carefully designed to teach the required electronics theory and *The FCC Rule Book* published by the ARRL you will soon have your Technician license. You will find the operating privileges available to you were worth the time spent learning about your hobby.

Until recently, the Novice license was the only entry route for most newcomers to Amateur Radio. The Novice license exam covers basic theory and regulations and a 5-word-per-minute (WPM) code test. Novices are allowed to operate Morse code on four high-frequency (HF) bands, where worldwide communication is possible, *without* satellites or repeater stations. Novices are also allowed to use voice on one HF band and two higher-frequency bands mostly used for local communications. Other privileges include radioteletype and computer-to-computer communications on the same bands where voice operation is permitted.

The Technician class license was usually the next step up for many radio amateurs. Until 1991, the Technician license required the same Morse code exam as the Novice license, and a slightly more difficult exam covering electronics and radio theory and Federal Communications Commission (FCC) rules and regulations. Technician licensees had the same operating privileges as Novices, *in addition* to full privileges on the amateur VHF and UHF bands.

Effective February 14, 1991, FCC removed the Morse code requirement for obtaining a Technician license. The theory and rules and regulations examination remained unchanged. So-called "code-free" Technician licensees have all the VHF and UHF privileges formerly granted to Technicians, but *not* the HF privileges granted to Novices. Those Novice-band privileges can be earned, however, by any code-free Technician licensee who passes a 5-WPM code test.

Any holder of a code-free Technician license who passes the code test is issued a *Certificate of Successful Completion of Examination* (CSCE) and is immediately granted Novice operating privileges on the HF bands. Novices who upgrade to Technician will still have their Novice HF privileges when they pass the Technician written exam. If you had a Novice license, but it expired less than two years ago, you will also be granted Novice HF privileges when you upgrade to Technician.

Technicians may operate on all frequency bands allocated to the amateur service above the HF bands (and there are a lot of them!). Table 1-2 shows the most popular bands Technicians can use. There are also amateur bands at the upper reaches of the frequency spectrum, where amateur work is largely experimental.

You can see that the Technician class license allows a wide range of transmitting privileges. To earn *all* those privileges, you will have to demonstrate that you know basic electronic theory, operating practices, and FCC rules and regulations. This book is designed to teach you the electronics and radio theory you need to know to pass the Element 3A license exam.

IF YOU'RE A NEWCOMER TO AMATEUR RADIO

Earning an Amateur Radio license, at whatever level, is a special achievement. The over half a million people in the US who call themselves Amateur Radio operators, or hams, are part of a global fraternity. Hams have fun operating two-way radio stations from their homes, cars and the out-of-doors. They talk with other hams across town and around the world using even the simplest of radio setups and antennas.

Just about anyone can be a ham: women, men, boys and girls. Hams range in age from under 8 to over 100! There are no limits. Hams come from all walks of life. Many famous people are hams, but the majority are ordinary folks who like making new friends every day. Nearly anyone can use Amateur Radio to open their door to the world.

Amateur Radio is primarily a hobby for personal enjoyment. But, in times of need, radio amateurs serve the public as a voluntary, noncommercial, communication service. This is especially true during natural disasters or other emergencies.

In addition, hams continue to make important contributions to the field of electronics. Amateur Radio experimentation is yet another reason many people become part of this self-disciplined group of trained operators, technicians and electronics experts—an asset to any country. Hams pursue their hobby purely for personal enrichment in technical and operating skills, without consideration of any type of payment.

Because radio signals do not know territorial boundaries, hams have a unique ability (and responsibility) to enhance international goodwill. As a ham, you become an ambassador of your country every time you put your station on the air.

Amateur Radio has been around since the early years of this century. Hams have always been at the forefront of technology. Today hams relay signals through their own satellites in the OSCAR (Orbiting Satellite Carrying Amateur Radio) series. They bounce signals off the moon and use any number of communications techniques. Amateurs talk from hand-held transceivers through mountaintop repeater stations. These repeaters relay their signals to transceivers in other hams' cars or homes. Hams send pictures by television, chat with other hams around the world by voice, "digital" modes or even Morse code. When emergencies arise, radio amateurs are on the spot to relay information to and from disaster-stricken areas that have lost normal lines of communication.

The US government, through the Federal Communications Commission (FCC), grants all US Amateur Radio licenses. This licensing procedure ensures that radio amateurs have the necessary operating skill and electronics knowledge. Without these skills, radio amateurs might unknowingly cause interference to other services using the radio spectrum because of improperly adjusted equipment or neglected regulations.

Who Can Be a Ham?

The FCC doesn't care how old you are or whether you're a US citizen: If you

Table 1-1

Amateur Operator Licenses†

Class	*Code Test*	*Written Examination*	*Privileges*
Novice	5 WPM (Element 1A)	Novice theory and regulations (Element 2)	Telegraphy on 3675-3725, 7100-7150 and 21,100-21,200 kHz with 200 watts PEP output maximum; telegraphy and RTTY on 28,100-28,300 kHz and telegraphy and SSB voice on 28,300-28,500 kHz with 200 W PEP max; all amateur modes authorized on 222.1-223.91 MHz, 25 W PEP max; all amateur modes authorized on 1270-1295 MHz, 5 W PEP max.
Technician		Novice theory and regulations; Technician-level theory and regulations. (Elements 2 and 3A)	All amateur privileges above 50.0 MHz. Technician-class licensees who have passed a 5-WPM code test also have Novice privileges.
General	13 WPM (Element 1B)	Novice theory and regulations; Technician and General theory and regulations. (Elements 2, 3A and 3B)	All amateur privileges except those reserved for Advanced and Amateur Extra class; see Table 1-2
Advanced	13 WPM (Element 1B)	All lower exam elements, plus Advanced theory. (Elements 2, 3A, 3B and 4A).	All amateur privileges except those reserved for Amateur Extra class, see Table 1-2.
Amateur Extra	20 WPM (Element 1C)	All lower exam elements plus Extra-class theory (Elements 2, 3A, 3B, 4A and 4B)	All amateur privileges

†A licensed radio amateur will be required to pass only those elements that are not included in the examination for the amateur license currently held.

pass the examination, the Commission will issue you an amateur license. Any person (except an agent of a foreign government) may take the exam, and, if successful receive an amateur license. It's important to understand that if a citizen of a foreign country receives an amateur license in this manner, he or she is a US Amateur Radio operator. (A citizen of another country, who already has an amateur license issued by that country, may be eligible for a reciprocal operating permit. Reciprocal operating agreements between the US and some foreign countries allow visitors from those countries to operate their own stations in the US without having to take an FCC exam.)

Licensing Structure

By examining Table 1-1, you'll see that there are five US amateur license

Table 1-2
Amateur Operating Privileges

160 METERS

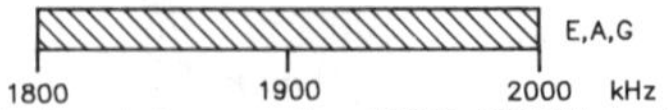

Amateur stations operating at 1900–2000 kHz must not cause interference to the radiolocation service and are afforded no protection from radiolocation operations.

80 METERS

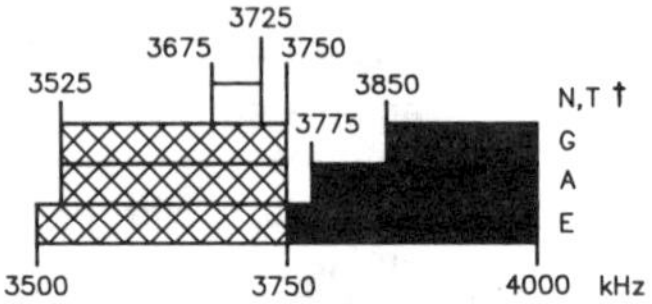

5,167.5 kHz (SSB only) Alaska emergency use only.

40 METERS

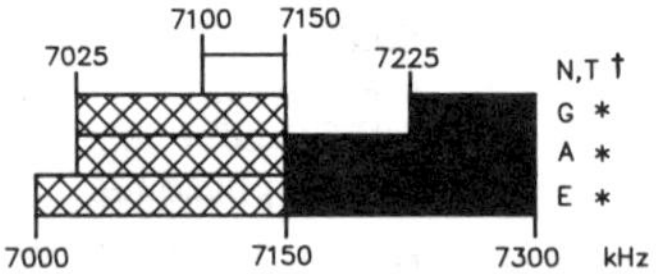

* Phone operation is allowed on 7075–7100 kHz in Puerto Rico; US Virgin Islands and areas of the Caribbean south of 20 degrees north latitude: and in Hawaii and areas near ITU Region 3, including Alaska.

30 METERS

Maximum power on 30 meters is 200 watts PEP output. Amateurs must avoid interference to the fixed service outside the US.

20 METERS

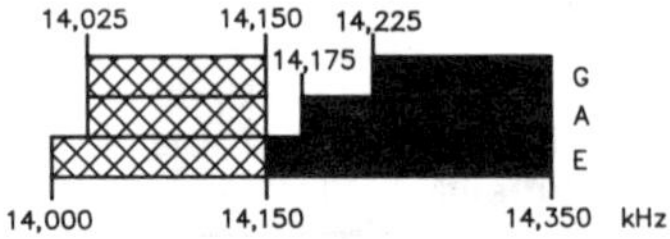

17 METERS

15 METERS

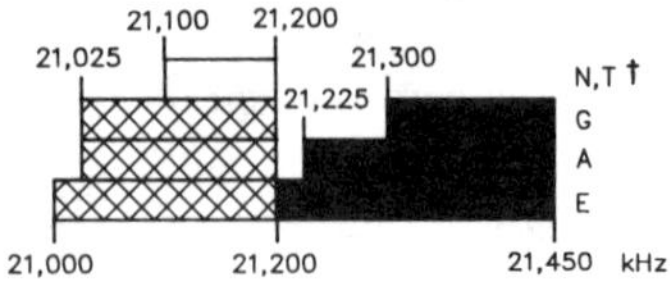

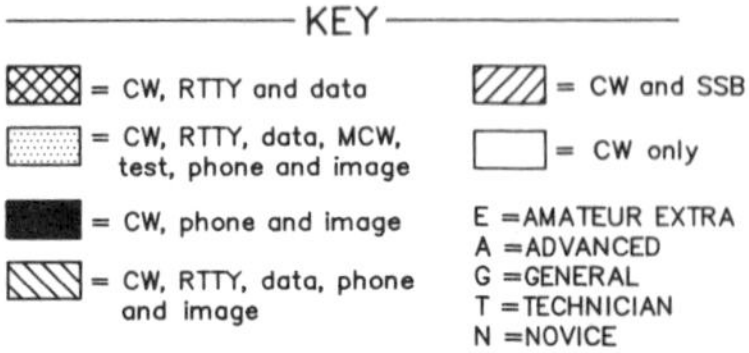

† Only Technician–class licensees who have passed a 5–WPM code test may use these frequencies.

** Geographical and power restrictions apply to these bands. See The FCC Rule Book for more information about your area.

12 METERS

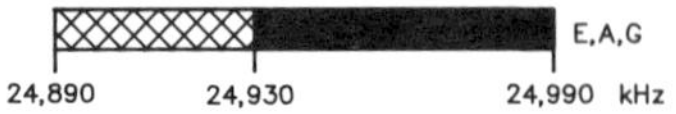

10 METERS

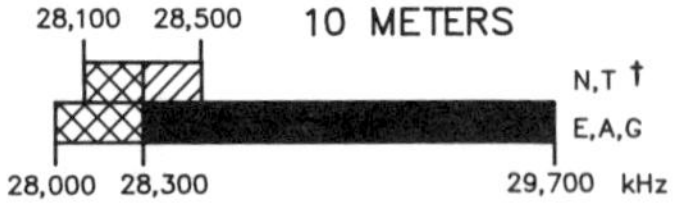

Novices and Technicians are limited to 200 watts PEP output on 10 meters.

6 METERS

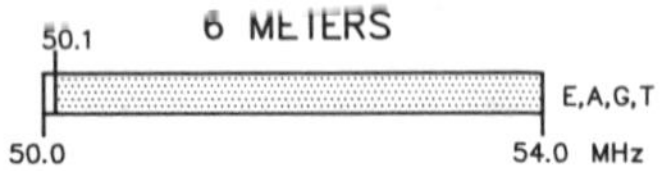

2 METERS

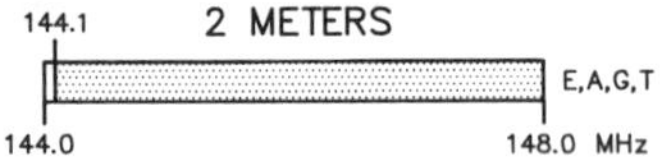

1.25 METERS

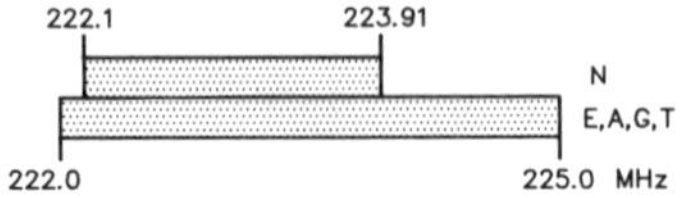

Novices are limited to 25 watts PEP output from 222.1 to 223.91 MHz.

70 CENTIMETERS **

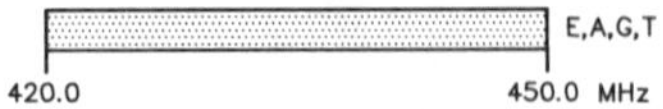

33 CENTIMETERS **

23 CENTIMETERS **

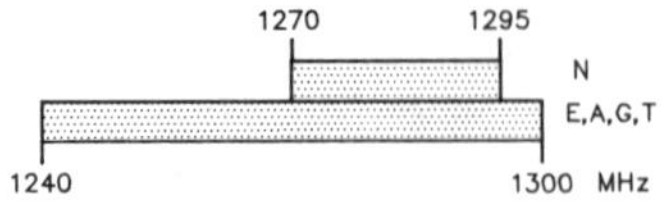

Novices are limited to 5 watts PEP output from 1270 to 1295 MHz.

classes. Each class has its own requirements and privileges. The FCC requires proof of your ability to operate an amateur station properly. The required knowledge is in line with the privileges of the license you hold. Higher license classes require more knowledge—and offer greater operating privileges. Specific operating privileges for Technician class licensees are listed in Table 1-2. As you upgrade your license class, you must pass more challenging written examinations.

In addition, you must show an ability to receive international Morse code at 5 words per minute (WPM) for Novice, 13 WPM for General and Advanced, and 20 WPM for Amateur Extra. (Every five letters count as a word. Numbers, punctuation and procedural signals count as two letters.) Although you may intend to use voice rather than code, you still must take the code test. By international treaty, knowing the international Morse code is a basic requirement for operating on any amateur band below 30 MHz.

The ARRL's package that teaches you the code, *Your Introduction to Morse Code*, includes two cassettes for beginners, and a booklet that will help you learn the code. Additional cassettes for code practice at speeds between 5 and 10 WPM, 10 and 15 WPM, and 15 and 22 WPM are available from the American Radio Relay League, Newington, CT 06111. These tapes use a character speed of 18 WPM with additional space between letters and words to slow the overall speed for practice at less than 18 WPM.

Code Practice

Besides listening to code tapes, on-the-air operating experience is a great help in building your code speed. When you have to copy the code the other station is sending to continue the conversation, your copying ability will improve quickly!

ARRL's Hiram Percy Maxim Memorial Station, W1AW, transmits Morse code practice and bulletins of interest to all amateurs. These code-practice sessions and bulletins provide an excellent opportunity for code practice. Table 1-3 is an abbreviated W1AW operating schedule. When we change from Standard Time to Daylight Savings Time, the same local times are used.

Station Call Signs

Many years ago most of the world's nations agreed to allocate certain call-sign prefixes to each country. This means that if you hear a radio station call sign beginning with K, W or N, for example, you know the station is licensed by the United States. A call sign beginning with the letter G is licensed by Great Britain, and so on.

International Telecommunication Union (ITU) radio regulations outline the basic principles used in forming amateur call signs. According to these regulations, an amateur call sign must be made of one or two characters (either of which may be a numeral) as a prefix. This is followed by a numeral and then a suffix of not more than three letters. The prefixes A (followed by another letter), K, N and W are used in the United States. The continental US is divided into 10 Amateur Radio call districts (or *call areas*), numbered 0 through 9. Figure 1-1 is a map showing the US call districts.

All US Amateur Radio call signs assigned by the FCC after March 1978 fit into one of five groups, each corresponding to a class, or classes, of license. Call signs are issued systematically by the FCC; requests for special call signs are not granted. For further information on the FCC's call-sign assignment system, and a table listing the blocks of call signs for each license class, see *The FCC Rule Book*, published by the ARRL.

Table 1-3

W1AW schedule

Pacific	Mtn	Cent	East	Sun	Mon	Tue	Wed	Thu	Fri	Sat
6 am	7 am	8 am	9 am			Fast Code	Slow Code	Fast Code	Slow Code	
7 am	8 am	9 am	10 am			Code Bulletin				
8 am	9 am	10 am	11 am			Teleprinter Bulletin				
9 am	10 am	11 am	noon							
10 am	11 am	noon	1 pm		Visiting Operator Time					
11 am	noon	1 pm	2 pm							
noon	1 pm	2 pm	3 pm							
1 pm	2 pm	3 pm	4 pm	Slow Code	Fast Code	Slow Code	Fast Code	Slow Code	Fast Code	Slow Code
2 pm	3 pm	4 pm	5 pm	Code Bulletin						
3 pm	4 pm	5 pm	6 pm	Teleprinter Bulletin						
4 pm	5 pm	6 pm	7 pm	Fast Code	Slow Code	Fast Code	Slow Code	Fast Code	Slow Code	Fast Code
5 pm	6 pm	7 pm	8 pm	Code Bulletin						
6 pm	7 pm	8 pm	9 pm	Teleprinter Bulletin						
6^{45} pm	7^{45} pm	8^{45} pm	9^{45} pm	Voice Bulletin						
7 pm	8 pm	9 pm	10 pm	Slow Code	Fast Code	Slow Code	Fast Code	Slow Code	Fast Code	Slow Code
8 pm	9 pm	10 pm	11 pm	Code Bulletin						
9 pm	10 pm	11 pm	Mdnte	Teleprinter Bulletin						
9^{45} pm	10^{45} pm	11^{45} pm	12^{45} am	Voice Bulletin						

❑ ***Morse code transmissions:***

Frequencies are 1.818, 3.5815, 7.0475, 14.0475, 18.0975, 21.0675, 28.0675 and 147.555 MHz.

Slow Code = practice sent at 5, 7½, 10, 13 and 15 WPM.

Fast Code = practice sent at 35, 30, 25, 20, 15, 13 and 10 WPM.

Code practice text is from the pages of *QST*. The source is given at the beginning of each practice session and alternate speeds within each session. For example, "Text is from July 1992 *QST*, pages 9 and 81," indicates that the plain text is from the article on page 9 and mixed number/letter groups are from page 81.

Code bulletins are sent at 18 WPM.

❑ ***Teleprinter transmissions:***

Frequencies are 3.625, 7.095, 14.095, 18.1025, 21.095, 28.095 and 147.555 MHz.

Bulletins are sent at 45.45-baud Baudot and 100-baud AMTOR, FEC Mode B.
110-baud ASCII will be sent only as time allows.

❑ ***Voice transmissions:***

Frequencies are 3.99, 7.29, 14.29, 18.16, 21.39, 28.59 and 147.555 MHz.

❑ ***Miscellanea:***

On Fridays, UTC, a DX bulletin replaces the regular bulletins.

W1AW is open to visitors during normal operating hours: from 1 PM until 1 AM on Mondays, 9 AM until 1 AM Tuesday through Friday, and from 3:30 PM to 1 AM on Saturdays and Sundays. FCC licensed amateurs may operate the station from 1-4 PM Monday through Friday. Be sure to bring your current FCC amateur license or a photocopy.

In a communications emergency, monitor W1AW for special bulletins as follows: voice on the hour, teleprinter at 15 minutes past the hour, and CW on the half hour.

Headquarters and W1AW are closed on New Year's Day, President's Day, Good Friday, Memorial Day, Independence Day, Labor Day, Thanksgiving and the following Friday, and Christmas Day. On the first Thursday of September, Headquarters and W1AW will be closed during the afternoon.

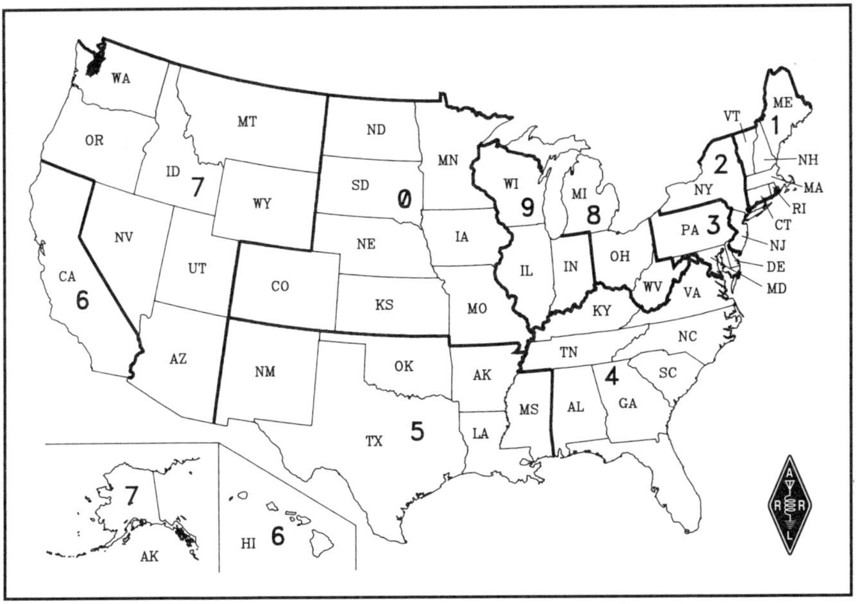

Figure 1-1—There are 10 US call areas. Hawaii is part of the sixth call area, and Alaska is part of the seventh.

If you already have an amateur call sign, you may keep the same one when you change license class if you wish. You must indicate your preference to receive a new call sign when you fill out an FCC Form 610 to apply for the exam or change your address.

EARNING A LICENSE

Applying for an Exam: FCC Form 610

Before you can take an FCC exam, you'll have to fill out a Form 610. This form is used as an application for a new license, a renewal of a license or a modification to a license.

Licenses are normally good for ten years. You should apply for a license renewal about 90 days before the license expires. If you forget to apply before it expires, you may still be able to renew without taking another exam. If you have a 10-year-term license, issued since January 1984, the grace period for reinstatement is two years.

Prior to January 1984, licenses were issued for five years. If you had one of those licenses but forgot to renew it, there is a grace period of five years after its expiration. If less than five years have elapsed since that license expired, you can still renew it without taking the exam again. After two years of the grace period elapsed, you lost your call sign and will be issued a new one.

If you apply for reinstatement of your license under the grace period, you may not operate your station until your reinstated ticket arrives.

To obtain a new Form 610, send a business-sized, self-addressed, stamped envelope to Federal Communications Commission, Forms Distribution Center, 2803

52nd Avenue, Hyattsville, MD 20781 or Form 610, ARRL, 225 Main St, Newington, CT 06111-1494.

Volunteer Examiner Program

All US amateur exams except the Novice are administered by volunteer examiners who are certified by a Volunteer Examiner Coordinator (VEC). *The FCC Rule Book* contains details on the volunteer examiner program.

To qualify for a Technician class license, you must pass FCC Exam Element 2 and Element 3A. If you already hold a valid Novice license, then you have credit for passing Elements 1A and 2, and will not have to retake them when you go for your Technician class exam. See Table 1-1 for details.

The Element 3A exam consists of 25 questions taken from a pool of more than 250. The FCC specifies subelements, or divisions, for each exam element. These are the Commission's Rules, Operating Procedures, Radio-Wave Propagation, Amateur Radio Practices, Electrical Principles, Circuit Components, Practical Circuits, Signals and Emissions, and Antennas and Feed Lines.

The question pools for all amateur exams are maintained by a Question Pool Committee selected by the Volunteer Examiner Coordinators. The FCC allows Volunteer Examiners to select the questions for an amateur exam, but they must use the questions exactly as they are released by the VEC that coordinates the test session. If you attend a test session coordinated by the ARRL/VEC, your test will be designed by the ARRL/VEC or by a computer program designed by the VEC, and the questions and answers will be exactly as they are printed in Chapter 10.

Finding an Exam Opportunity

To determine where and when an exam will be given, contact the ARRL/VEC Office, or watch for announcements in the Hamfest Calendar and Coming Conventions columns in *QST*. Many local clubs sponsor exams, so they are another good source of information on exam opportunities. ARRL officials such as Directors, Vice Directors and Section Managers receive notices about test sessions in their area. See page 9 in the latest issue of *QST* for their names and addresses.

To register for an exam, send a completed Form 610 to the Volunteer Examiner Team responsible for the exam session if preregistration is required. Otherwise, bring the form to the session. Registration deadlines, and the time and location of the exams, are mentioned prominently in publicity releases about upcoming sessions.

Taking the Exam

By the time examination day rolls around, you should have already prepared yourself. This means getting your schedule, supplies and mental attitude ready. Plan your schedule so you'll get to the examination site with plenty of time to spare. There's no harm in being early. In fact, you might have time to discuss hamming with another applicant, which is a great way to calm pretest nerves. Try not to discuss the material that will be on the examination, as this may make you even more nervous. By this time, it's too late to study anyway!

What supplies will you need? First, be sure you bring your current *original* Amateur Radio license, if you have one. Bring along several sharpened number 2 pencils and two pens (blue or black ink). Be sure to have a good eraser. A pocket calculator may also come in handy. You may use a programmable calculator if that is the kind you have, but take it into your exam "empty" (cleared of all programs and constants in memory). Don't program equations ahead of time because you may be asked to demonstrate there is nothing in the calculator's memory.

The Volunteer Examiner Team is required to check two forms of identification before you enter the test room. This includes your *original* Amateur Radio license. A photo ID of some type is best for the second form, but is not required by the FCC. Other acceptable forms of identification include a driver's license, a piece of mail addressed to you, a birth certificate or some other such document.

The following description of the testing procedure applies to exams coordinated by the ARRL/VEC, although many other VECs use a similar procedure.

Code Tests

Since you already have a Novice license, you have credit for passing the 5 WPM Element 1A Morse-code test. You may want to try for the 13 WPM code exam (Element 1B), in which case this information will be of interest to you. Even if you aren't planning to take a code exam this time, you should be prepared to sit through the code exams for those who are taking them.

The code tests are usually given before the written exams. The 20 WPM exam normally is given first, then the 13 WPM exam and finally the 5 WPM test. There is no harm in trying the 20 or 13 WPM exam, even though you are testing for the Technician license.

Before you take the code test, you'll be handed a piece of paper to copy the code as it's sent. The test will begin with about a minute of practice copy. Then comes the actual test: five minutes of Morse code. You are responsible for knowing the 26 letters of the alphabet, the numerals 0 through 9, the period, comma, question mark and the procedural signals $\overline{\text{AR}}$, $\overline{\text{SK}}$, $\overline{\text{BT}}$ and $\overline{\text{DN}}$. You may copy the entire text word for word, or just take notes on the content. At the end of the transmission, the examiner will hand you 10 questions about the text. Simply fill in the blanks with your answers. (You must spell each answer exactly as it was sent.) Some Volunteer Examiner Teams are using multiple-choice code exams. If you get at least 7 correct, you pass! Alternatively, the exam team has the option to look at your copy sheet if you fail the 10-question exam. If you have one minute of copy with no mistakes, they can certify that you passed the test on that basis. The format of the test transmission is similar to one side of a normal on-the-air amateur conversation.

A sending test may not be required. The Commission has decided that if applicants can demonstrate receiving ability, they most likely can also send at that speed. But be prepared for a sending test, just in case! Subpart 97.503(a) of the FCC rules says, "A telegraphy examination must be sufficient to prove that the examinee has the ability to send correctly by hand and to receive correctly by ear texts in the international Morse code at not less than the prescribed speed . . ."

Written Tests

After the code tests are administered, you'll take the written examination. The examiner will give each applicant a test booklet, an answer sheet and scratch paper. After that, you're on your own. The first thing to do is read the instructions. Be sure to sign your name every place it's called for. Do all of this at the beginning to get it out of the way.

Next, check the examination to see that all pages and questions are there. If not, report this to the examiner immediately. When filling in your answer sheet, make sure your answers are marked next to the numbers that correspond to each question.

Go through the entire exam, and answer the easy questions first. Next, go back to the beginning and try the harder questions. Leave the really tough questions for last. Guessing can only help, as there is no additional penalty for answering incorrectly.

If you have to guess, do it intelligently: At first glance, you may find that you can eliminate one or more "distracters." Of the remaining responses, more than one may seem correct; only one is the best answer, however. To the applicant who is fully prepared, incorrect distracters to each question are obvious. Nothing beats preparation!

After you've finished, check the examination thoroughly. You may have read a question wrong or goofed in your arithmetic. Don't be overconfident. There's no rush, so take your time. Think, and check your answer sheet. When you feel you've done your best and can do no more, return the test booklet, answer sheet and scratch paper to the examiner.

The Volunteer Examiner Team will grade the exam right away. The passing mark is 74%. (That means no more than 6 incorrect answers.) You will receive a *Certificate of Successful Completion of Examination* (CSCE) showing all exam elements that you pass. If you are already licensed, and you pass the exam elements required to earn a higher class of license, the CSCE authorizes you to operate with your new privileges. When you use these new privileges, you must sign your call sign, followed by the slant mark ("/"; on voice, say "stroke" or "slant") and the letters KT (for Novices upgrading to Technician). There is one exception to this rule: If you had a code-free Technician license and pass the 5-WPM code test, you do not have to use the "KT" indicator when you use your new privileges or when you operate on the bands already allocated to Technicians (50 MHz and above).

If you pass only some of the exam elements required for a higher class license you will still receive a CSCE. That certificate shows what exam elements you passed, and is valid for one year. Use it as proof that you passed those exam elements so you won't have to take them over again next time you try for the upgrade.

AND NOW, LET'S BEGIN

Subelement T1 in Chapter 10 covers the rules and regulations for the Element 3A exam. You should use *The FCC Rule Book* to find the material covered by those questions. Then go over that section of the question pool to check your understanding of the rules. Perhaps you will want to study the rules a few at a time, using that as a break from your study in the rest of this book.

There you have it. The remainder of this book provides the background in electronics theory that you will need to pass the Element 3A Technician class written exam. Table 1-4 shows the study guide or syllabus for the Element 3A exam. This study guide was released by the Volunteer Examiner Coordinators' Question Pool Committee in 1992. The syllabus lists the topics to be covered by the Technician exam, and so forms the basic outline for the remainder of this book. Use the syllabus to guide your study, and to ensure that you have studied the material for all of the topics listed.

The question numbers used in the question pool refer to this syllabus. Each question number begins with a syllabus-point number (T1C, T2A, T3B). The question numbers end with a two digit number. For example, question T1C02 is the second question in the series about syllabus point T1C. Question T3B09 is the ninth question about the T3B syllabus point.

The Question Pool Committee designed the syllabus and question pool so there are the same number of points in each subelement as there are exam questions from that subelement. For example, the FCC specifies 5 questions on the Technician exam must come from the "Commission's Rules" subelement. There are five groups of questions for this subelement, T1A through T1E. Three exam questions must be from the "Antennas and Feed Lines" subelement, so there are three groups for that point. These are numbered T9A, T9B and T9C.

Good luck with your studies!

Table 1-4

Element 3A (Technician) Syllabus

(Required for all operator licenses except Novice)

SUBELEMENT T1 COMMISSION'S RULES
[5 exam questions—5 groups]

T1A Station control, frequency privileges authorized to the Technician class control operator, term of licenses, grace periods and modifications of licenses.

T1B Emission privileges for Technician class control operator, frequency selection and sharing, transmitter power.

T1C Digital communications, station identification, ID with CSCE.

T1D Correct language, phonetics, beacons and radio control of model craft and vehicles.

T1E Emergency communications; broadcasting; permissible one-way, satellite and third party communication; indecent and profane language.

SUBELEMENT T2 OPERATING PROCEDURES
[3 exam questions—3 groups]

T2A Repeater operation, courteous operation.

T2B Simplex operations, Q signals, RST signal reporting, repeater frequency coordination.

T2C Distress calling and emergency drills and communications—operations and equipment, Radio Amateur Civil Emergency Service (RACES).

SUBELEMENT T3 RADIO-WAVE PROPAGATION
[3 exam questions—3 groups]

T3A Ionosphere, ionospheric regions, solar radiation.

T3B Ionospheric absorption, causes and variation, maximum usable frequency.

T3C Propagation, including ionospheric, tropospheric, line-of-sight, scatter propagation, and maximum usable frequency.

SUBELEMENT T4 AMATEUR RADIO PRACTICES
[4 Exam Questions—4 groups]

T4A Electrical wiring, including switch location, dangerous voltages and currents.

T4B Meters, including volt, amp, multi, peak-reading, RF watt and placement, and ratings of fuses and switches.

T4C Marker generator, crystal calibrator, signal generators and impedance-match indicator.

T4D Dummy antennas, S-meter, exposure of the human body to RF.

SUBELEMENT T5 ELECTRICAL PRINCIPLES
[2 exam questions—2 groups]

T5A Definition of resistance, inductance, and capacitance and unit of measurement, calculation of values in series and parallel.

T5B Ohm's Law.

SUBELEMENT T6 CIRCUIT COMPONENTS
[2 Exam Questions—2 groups]

T6A Resistors, construction types, variable and fixed, color code, power ratings, schematic symbols.

T6B Schematic symbols—inductors and capacitors, construction of variable and fixed, factors affecting inductance and capacitance, capacitor construction.

SUBELEMENT T7 PRACTICAL CIRCUITS
[1 exam question—1 group]

T7A Practical circuits.

SUBELEMENT T8 SIGNALS AND EMISSIONS
[2 exam questions—2 groups]

T8A Definition of modulation and emission types.

T8B RF carrier, modulation, bandwidth and deviation.

SUBELEMENT T9 ANTENNAS AND FEED LINES
[3 exam questions—3 groups]

T9A Parasitic beam and non-directional antennas.

T9B Polarization, impedance matching and SWR, feed lines, balanced vs unbalanced (including baluns).

T9C Line losses by line type, length and frequency, RF safety.

KEY WORDS

Break—An interruption of a QSO. When said twice over a repeater, it means "I must make a distress call immediately!"

Closed—A repeater that restricts access to those who know a special code.

Courtesy tone—A sound heard on a repeater indicating that a transmission is complete.

CW—A communications mode transmitted by on/off keying of a radio-frequency signal. Another name for international Morse code.

Dummy load—A station accessory that allows you to test or adjust transmitting equipment without sending a signal out over the air. Also called *dummy antenna*.

Duplex operation—Receiving and transmitting on two different frequencies. Also see **simplex operation**.

Emergency—A situation where there is a danger to lives or property.

Emergency traffic—Messages concerning the immediate safety of human life or the immediate protection of property.

Frequency coordinator—An individual or group that recommends repeater frequencies.

Frequency separation—The difference between a repeater's receive and transmit frequencies.

Frequency-modulated (FM) voice—Mode of voice (phone) communications used on repeaters.

Health-and-Welfare traffic—Messages concerning a person's well being.

Input frequency—A repeater's receiving frequency. To use a repeater, transmit on the input frequency and receive on the **output frequency**.

MAYDAY—The distress call used on phone.

Morse code—(See **CW**.)

Open—A repeater that can be used by all properly licensed hams.

Output frequency—A repeater's transmitting frequency. To use a repeater, transmit on the **input frequency** and receive on the output frequency.

Phone—Another name for voice communications, such as SSB and FM.

Q signal—Three-letter symbol beginning with Q. Used on CW to save time and to improve communication. Some examples are QRS (send slower), QTH (location), QSO (ham conversation) and QSL (acknowledgment of receipt).

QRL?—Ham radio Q signal meaning "Is this frequency in use?"

QRZ?—Ham radio Q signal meaning "Is someone calling me?"

QSO—A conversation between two radio amateurs.

QSY—Ham radio Q signal meaning "change frequency."

Quieting—On repeaters, signal reports are given in terms of the amount of quieting. *Full quieting* means the signal is strong enough to overcome all receiver noise.

RACES (Radio Amateur Civil Emergency Service)—A part of the amateur service that provides radio communications only for civil defense purposes.

Repeater—An amateur station that receives a signal and retransmits it for greater range.

RST—A system of numbers used for signal reports: R is readability, S is strength and T is tone. (On single-sideband phone, only R and S reports are used.)

Signal report—An exchange between stations indicating the relative strength of each other's signals.

Simplex operation—Receiving and transmitting on the same frequency. See **duplex operation**.

Single sideband (SSB)—Emission mode that describes the type of voice emission used most on the HF bands.

$\overline{\text{SOS}}$—The distress call used on CW.

Tactical call signs—Type of station identification used to help coordinate public-service events. Examples are *parade headquarters* and *net control*.

Time-out timer—A device that limits the amount of time any one person can talk through a repeater.

2 Operating Procedures

In this chapter, we'll give you some tips for operating with your Technician license—on CW, on SSB phone and on FM. Then we'll take a look at emergency procedures and drills.

A good rule for Amateur Radio contacts (**QSOs**) is to talk as you would during a face-to-face conversation. When you meet someone for the first time, you introduce yourself once; you don't repeat your name over and over again. An exception would be if the other person is hard of hearing or you're meeting in a noisy place. Even then you would only repeat if the person couldn't hear you. When you are on the air, the other operator will tell you if she can't hear you clearly. Problem situations on HF include static (QRN), interference (QRM) or signal fading (QSB). On voice, say what you mean.

Before getting on the air, keep in mind that you should never transmit a signal over the air when you're testing or adjusting an HF rig. If your transmitter requires tuning when you change bands, connect a **dummy load** to your transmitter output while you tune. You should minimize the time you transmit over the air when tuning your radio. Using a dummy load will reduce the possibility that you will cause on-the-air interference. (Chapter 4 has more information on dummy loads.) Once you have tuned up the transceiver according to the instruction manual, disconnect the dummy load and connect the antenna. An antenna switch makes this easy. Now, you're ready to operate!

CW Q SIGNALS AND SIGNAL REPORTS

Many hams use international **Morse code**, or **CW**, to communicate; some will use nothing else. It's efficient, it gets through when voice modes won't and it's fun.

CW operators use **Q signals** to save time and increase efficiency. It's far easier to send, for example, **QRZ?** than IS ANYONE CALLING ME? A couple of other common Q signals are **QRL?** (Is this frequency in use?), **QSO** (ham conversation) and **QSY** (change frequency). Table 2-1 lists other common Q signals. Although hams use some Q signals in face-to-face conversations, when you're on the air, use them on CW only.

Another CW shortcut is to use the **RST** (*readability*, *strength* and *tone*) system of reporting signal reception. You'll exchange a **signal report** in nearly every Amateur Radio QSO. The scales are a general indication of how you are receiving the other station. As you gain experience with the descriptions given in Table 2-2, you'll be more comfortable estimating signal reports.

A report of RST 368 would be interpreted as "Your signal is readable with considerable difficulty, good strength, with a slight trace of modulation."

Table 2-1

Q Signals

These Q signals are the ones used most often on the air. (Q abbreviations take the form of questions only when they are sent followed by a question mark.)

QRG	Will you tell me my exact frequency (or that of ___)? Your exact frequency (or that of ___) is ___ kHz.
QRL	Are you busy? I am busy (or I am busy with ___). Please do not interfere.
QRM	Is my transmission being interfered with? Your transmission is being interfered with ___ (1. Nil; 2. Slightly; 3. Moderately; 4. Severely; 5. Extremely.)
QRN	Are you troubled by static? I am troubled by static ___. (1-5 as under QRM).
QRO	Shall I increase power? Increase power.
QRP	Shall I decrease power? Decrease power.
QRQ	Shall I send faster? Send faster (___ WPM).
QRS	Shall I send more slowly? Send more slowly (___ WPM).
QRT	Shall I stop sending? Stop sending.
QRU	Have you anything for me? I have nothing for you.
QRV	Are you ready? I am ready.
QRX	When will you call me again? I will call you again at ___ hours (on ___ kHz).
QRZ	Who is calling me? You are being called by ___(on ___kHz).
QSB	Are my signals fading? Your signals are fading.
QSK	Can you hear me between your signals and if so can I break in on your transmission? I can hear you between signals; break in on my transmission.
QSL	Can you acknowledge receipt (of a message or transmission)? I am acknowledging receipt.
QSN	Did you hear me (or ___) on ___kHz? I did hear you (or ___) on ___kHz.
QSO	Can you communicate with ___direct or by relay? I can communicate with ___ direct (or relay through ___).
QSP	Will you relay to ___? I will relay to ___.
QST	General call preceding a message addressed to all amateurs and ARRL members. This is in effect "CQ ARRL."
QSX	Will you listen to ___on ___kHz? I am listening to ______ on ___kHz.
QSY	Shall I change to transmission on another frequency? Change to transmission on another frequency (or on ___ kHz).
QTB	Do you agree with my counting of words? I do not agree with your counting of words. I will repeat the first letter or digit of each word or group.
QTC	How many messages have you to send? I have ___ messages for you (or for ___).
QTH	What is your location? My location is ___
QTR	What is the correct time? The time is ___

Table 2-2

The RST System

READABILITY
1—Unreadable.
2—Barely readable, occasional words distinguishable.
3—Readable with considerable difficulty.
4—Readable with practically no difficulty.
5—Perfectly readable.

SIGNAL STRENGTH
1—Faint signals barely perceptible.
2—Very weak signals.
3—Weak signals.
4—Fair signals.
5—Fairly good signals.
6—Good signals.
7—Moderately strong signals.
8—Strong signals.
9—Extremely strong signals.

TONE
1—Sixty-cycle ac or less, very rough and broad.
2—Very rough ac, very harsh and broad.
3—Rough ac tone, rectified but not filtered.
4—Rough note, some trace of filtering.
5—Filtered rectified ac but strongly ripple-modulated.
6—Filtered tone, definite trace of ripple modulation.
7—Near pure tone, trace of ripple modulation.
8—Near perfect tone, slight trace of modulation.
9—Perfect tone, no trace of ripple or modulation of any kind.

The "tone" report refers only to the purity of the signal. It has no connection with its stability or freedom from clicks or chirps. Most of the signals you hear will be a T-9. Other tone reports occur mainly if the power supply filter capacitors are not doing a thorough job. If so, some trace of ac ripple finds its way onto the transmitted signal. If the signal has the characteristic steadiness of crystal control, add X to the report (for example, RST 469X). If it has a chirp or "tail" (either on "make" or "break") add C (for example, 469C). If it has clicks or noticeable other keying transients, add K (for example, 469K). Of course a signal could have both chirps and clicks, in which case both C and K could be used (for example, RST 469CK).

SIGNAL REPORTS ON SSB PHONE AND FM REPEATERS

On **single sideband (SSB phone)**, signal reports are two-digit numbers using the RS portion of the RST system. No tone report is required. The maximum signal report would be "59"—that is, readability 5 (perfectly readable), strength 9 (extremely strong signals). If you get a report of, say, "59 plus 20 dB," it means your received signal reads 20 decibels higher than signal strength 9 on the transceiver S meter. A report of "57" means perfectly readable and moderately strong. A report of "33," on the other hand, means readable with considerable difficulty and weak. On FM repeaters, RS reports are not appropriate. FM signal reports are generally given as the amount of **quieting**. *Full quieting* means the received signal is strong enough to block all receiver noise.

[Now it's time to turn to Chapter 10 to review questions T2A09, T2A10, T2A17 and T2B05 through T2B11. Review this section if you have any difficulty.]

VHF AND UHF OPERATING

Operating on VHF and UHF bands takes many forms. You can use SSB and CW for *weak-signal* operation, you can use amateur satellites and you can use **frequency-modulated phone (FM voice)** on simplex and repeaters. Before we look at repeaters in detail, a brief reminder: It's a good idea to use VHF or UHF for all local communications. The HF bands should be reserved for longer-distance contacts to reduce interference on those bands.

VHF and UHF FM voice operation takes two forms: simplex and repeater. **Simplex operation** means the stations are talking to each other directly, on the same frequency. This is similar to making a contact on the HF bands.

The communications range for VHF and UHF FM simplex is usually limited to your local area. If you live high on a mountain and use a high-gain directional antenna, you may be able to extend your range considerably. Unfortunately, most of us do not have the luxury of ideal VHF/UHF operating conditions. Often, we want to make contacts even though we live in a valley, are driving in a car or are using a low-power, hand-held transceiver. Enter repeaters.

Fm and Repeaters

A **repeater** receives a signal on one frequency and simultaneously retransmits (repeats) it on another frequency. Receiving and transmitting on two different frequencies is called **duplex operation**. The frequency it receives on is called the **input frequency**, and the frequency it transmits on is called the **output frequency**. Often located atop a tall building or high mountain, VHF and UHF repeaters greatly extend the operating range of amateurs using mobile and hand-held transceivers. See Figure 2-1.

Equipment Considerations

To use a repeater, you must have a transceiver that can transmit on the repeater's input frequency—the frequency the repeater receives on. Your radio must also be able to receive on the repeater output frequency—the frequency the repeater transmits on. The input and output frequencies are separated by a predetermined amount that is different for each band. This **frequency separation** is called the *offset*. Common repeater offsets for the VHF and UHF bands are listed in Table 2-3. For example, the offset on 1.25 meters is 1.6 MHz. A repeater on 1.25 meters might have its input frequency on 222.32 MHz and its output on 223.92 MHz.

Most repeaters are **open**—that is, available for use by anyone in range. Some repeaters, however, are restricted to certain groups, such as members of a club. Such **closed** repeaters require the transmission of a continuous subaudible tone or a short

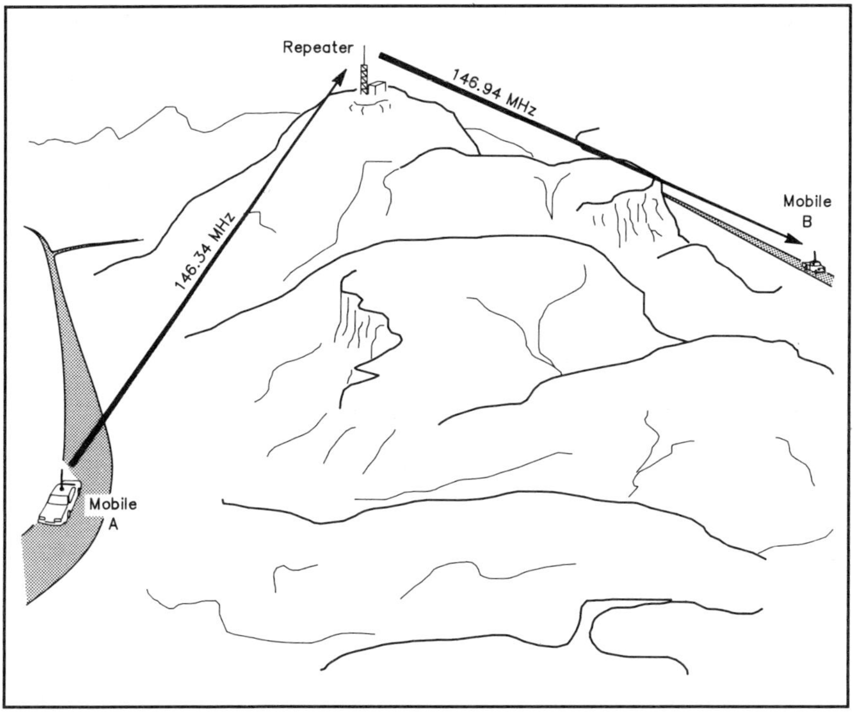

Figure 2-1—The mountain blocks direct (simplex) communication between mobile stations A and B. When mobile A transmits on 146.34 MHz, however, the repeater detects its signal and relays it to mobile B on 146.94 MHz. When it's mobile B's turn to transmit, the same signal relay occurs in reverse.

"burst" of tones for access. These are called *CTCSS* (continuous tone-coded squelch system) or *PL* tones. Some open repeaters also require the use of special codes or subaudible tones to gain access. This prevents interference from extraneous transmissions that might accidentally key the repeater. If you wish to join a group that sponsors a closed repeater, contact the repeater control operator.

Repeaters can be found on all VHF and UHF bands. Repeater frequencies are selected through consultation with **frequency coordinators**—individuals or groups that recommend repeater frequencies based on potential interference and other factors.

Table 2-3

Repeater Frequency-Separation Offsets

Band	*Offset*
6 meters	1 MHz
2 meters	600 kHz
1.25 meters	1.6 MHz
70 cm	5 MHz
33 cm	12 MHz
23 cm	20 MHz

Repeater Operating

Before you make your first FM repeater contact, you should learn some repeater operating techniques. It's worth a few minutes to listen and familiarize yourself with the procedures used by other hams in your area. Accepted procedures can vary slightly from repeater to repeater.

Your First Transmission

Making your first transmission on a repeater is as simple as signing your call. If the repeater is quiet, just say "KA4XYZ" or "KA4XYZ listening"—to attract someone's attention. After you stop transmitting, you will usually hear the unmodulated repeater carrier for a second or two. This *squelch tail* lets you know that the repeater is working. Someone interested in talking to you will call you after your initial transmission. Some repeaters have specific rules for making yourself heard. In general, however, your call sign is all you need.

Don't call CQ to initiate a conversation. It takes longer to complete a CQ than to transmit your call sign. (In some areas, a solitary "CQ" is permissible.) Efficient communication is the goal. You are not on HF, trying to attract the attention of someone who is casually tuning across the band. In the FM mode, stations are either monitoring their favorite frequency or not. Except for scanner operation, there is not much tuning across the repeater bands.

To join a conversation in progress, transmit your call sign during a break between transmissions. The station that transmits next will usually acknowledge you. Don't use the word "break" to join a conversation—unless you want to use the repeater to help in an **emergency**.

To call another station when the repeater is not in use, just give both calls. For example, "WA1MBK, this is N1BKE." If the repeater is in use, but the conversation sounds like it is about to end, wait before calling another station. If the conversation sounds like it is going to continue for a while, however, transmit only your call sign between their transmissions. After you are acknowledged, ask to make a quick call. Usually, the other stations will stand by. Make your call short. If your friend responds, try to meet on another repeater or on a simplex frequency. Otherwise, ask your friend to stand by until the present conversation ends.

Use plain language on a repeater. If you want to know someone's location, say "Where are you?" If you want to know whether someone you're talking with is using a mobile rig or a hand-held radio, just ask: "What kind of radio are you using?" You get the idea.

Courtesy Counts

If you are in the midst of a conversation and another station transmits his or her call sign between transmissions, the next station in line to transmit should acknowledge the new station and permit the new arrival to make a call or join the conversation. It is impolite not to acknowledge new stations, or to acknowledge them but not let them speak. The calling station may need to use the repeater immediately. He or she may have an emergency to handle, so let him or her make a transmission promptly.

A brief pause before you begin each transmission allows other stations to break in—there could be an emergency. Don't key your microphone as soon as someone else releases hers. If your exchanges are too quick, you can prevent other stations from getting in.

The **courtesy tones** found on some repeaters force users to leave a space between transmissions. The beeper sounds a second or two after each transmission to permit new stations to transmit their call signs in the intervening time period. The conversation may continue only after the beeper sounds. If a station is too quick and begins transmitting before the beeper sounds, the repeater may indicate the violation, sometimes by shutting down!

Keep transmissions as short as possible, so more people can use the repeater. Again, long transmissions could prevent someone with an emergency from getting the chance to call for help through the repeater. All repeaters encourage short transmissions by "timing out" (shutting down for a few minutes) when someone gets long-winded. The **time-out timer** also prevents the repeater from transmitting continuously, due to distant signals or interference. Because it has such a wide coverage area, a continuously transmitting repeater could cause unnecessary interference. Continuous operation can also damage the repeater.

You must transmit your call sign at the end of a contact and at least every 10 minutes during the course of any communication. You do not have to transmit the call sign of the station to whom you are transmitting.

Never transmit without identifying. For example, keying your microphone to turn on the repeater without saying your station call sign is illegal. If you do not want to engage in conversation, but simply want to check if you are able to access a particular repeater, simply say "KJ4KB testing."

Fixed Stations and Prime Time

Repeaters were originally intended to enhance mobile communications. During commuter rush hours, mobile stations still have preference over fixed stations on some repeaters. During mobile prime time, fixed stations should generally yield to mobile stations. When you're operating as a fixed station, don't abandon the repeater completely, though. Monitor the mobiles; your assistance may be needed in an emergency. Use good judgment: Rush hours are not the time to test your radio extensively or to join a net that doesn't deal with the weather, highway conditions or other subjects related to commuting.

Simplex Operation

After you have made a contact on a repeater, move the conversation to a *simplex* frequency if possible. The repeater is not a soapbox. You may like to listen to yourself, but others, who may need to use the repeater, will not appreciate your tieing up the repeater unnecessarily.

The function of a repeater is to provide communications between stations that can't otherwise communicate because of terrain, equipment limitations or both. It follows that stations able to communicate without a repeater should not use one. That way, the repeater is available for stations that need it. (Besides, communication on simplex offers a degree of privacy impossible to achieve on a repeater. On simplex you can usually have extensive conversations without interruption.)

Select a frequency designated for FM simplex operation. Otherwise, you may interfere with stations operating in other modes without realizing it. (The reason for this is simple: Changing to a simplex frequency is far easier than changing the frequencies a repeater uses.) Each band has a designated national FM simplex calling frequency, which is the center for most simplex operation. To see if you and the other station can communicate on a simplex frequency, listen on the repeater input frequency. If you can clearly hear what's going into the repeater, you don't need the repeater to communicate.

[Time to review some questions in Chapter 10: T2A01 through T2A08, T2A11 through T2A16, T2A18, T2B01 through T2B04 and T8A06. Reread this section if you have any difficulty.]

EMERGENCY COMMUNICATIONS

Distress Calls

If you should require immediate emergency help, and you're using a voice (telephony) mode, call MAYDAY several times. Use whatever frequency offers the best chance of getting a useful answer. "MAYDAY" is from the French *m'aidez* (help me). On CW (telegraphy), use $\overline{\text{SOS}}$ to call for help. To make a distress call over a repeater, say "**break break**" and then your call sign to alert all stations to stand by while you deal with the emergency.

If necessary, repeat your distress call a few times, and pause for any station to answer. Identify the transmission with your call sign. Stations that hear the call sign will realize the distress call is legitimate. In a life- or property-threatening emergency, you may send a distress call on *any* frequency, even outside the amateur bands, if you think doing so will bring help faster.

Be ready to supply the following information to the stations responding to a distress call:

- *The location of the emergency*, with enough detail to permit rescuers to locate it without difficulty.
- *The nature of the distress situation.*
- *The type of assistance required* (medical, evacuation, food, clothing or other aid).
- *Any other information* to help locate the emergency area.

If you receive a distress signal, you are also allowed to transmit on *any* frequency to provide assistance.

RACES

RACES (Radio Amateur Civil Emergency Service) is a part of the amateur service that provides radio communications only for civil defense purposes. It is active *only* during periods of local, regional or national civil emergencies.

You must be registered with the responsible civil defense organization to operate as a RACES station. RACES stations may not communicate with amateurs not operating in a RACES capacity. Restrictions do not apply when stations are operating in a non-RACES amateur capacity, such as the ARES, the Amateur Radio Emergency Service.

Only civil-preparedness communications can be transmitted during RACES operation. These are defined in section 97.407 of the FCC regulations. Rules permit tests and drills for a maximum of one hour per week. All test and drill messages must be clearly identified as such.

[Now study questions T2C01 through T2C06 in Chapter 10. Review as needed.]

Tactical Communications

Tactical communications is first-response communications in an emergency involving a few people in a small area. This type of communications is unformatted and seldom written. It may be urgent instructions or requests such as "Send an ambulance," or "Who has the medical supplies?"

Tactical communications usually use 2-meter repeater net frequencies or the 146.52-MHz simplex calling frequency. Compatible mobile, portable and fixed-station equipment is plentiful and popular for these frequencies. Tactical communications is particularly important when working with local government and law-enforcement agencies. Use the 12-hour local-time system for times and dates when working with relief agencies. Most may not understand the 24-hour system or Coordinated Universal Time (UTC).

Another way to make tactical communications efficient is to use **tactical call signs**, which describe a function, location or agency. Their use promotes coordination with individuals or agencies who are monitoring. When operators change shifts or locations, the set of tactical call signs remains the same. Amateurs may use such tactical call signs as *parade headquarters*, *finish line*, *Red Cross* or *Net Control*. This procedure promotes efficiency and coordination in public-service communication activities. Tactical call signs do not fulfill the identification requirements of section 97.119 of the FCC rules, however. Amateurs must also identify their station operation with their FCC-assigned call sign. Identify at the end of the operation and at intervals not to exceed 10 minutes during operation.

Health-and-Welfare Traffic

There can be a large amount of radio traffic to handle during a disaster. Phone lines still in working order are often overloaded. They should be reserved for emergency use by those people in peril. Shortly after a major disaster, **Emergency traffic** leaves the disaster area. These have life-and-death urgency or are for medical help and critical supplies. Handle them first. Next is *Priority* traffic. These are emergency-related messages, but not as important as Emergency traffic. Then, handle **Health-and-Welfare traffic**, which pertains to the well being of evacuees or the injured. This results in timely advisories to those waiting outside the disaster area.

Hams inside disaster areas cannot immediately find out about someone's relative when their own lives may be threatened—they are busy handling emergency messages. After the immediate emergency subsides, concerned friends and relatives of possible victims can send Health-and-Welfare inquiries into the disaster area.

Emergency Equipment

When cities, towns, counties, states or the federal government call for emergency communications, amateurs answer. They press their mobile and portable radio equipment into service, using alternatives to commercial power. Generators, car batteries, windmills or solar energy can provide power for equipment during an emergency, when normal ac power is often out of service. A dipole antenna is the best choice for an HF station in emergencies—it can be installed easily, and wire is light and portable. Carry an ample supply of wire and you'll be ready to go on the air at any time and any place. For VHF/UHF operation, a groundplane antenna that you can string in a tree works well. If you're planning to use a battery-powered hand-held transceiver, bring along one or more spare, *charged* battery packs.

[Before going on to the next chapter, turn to Chapter 10 and study questions T2C07 through T2C12. Review this section if you have difficulty.]

KEY WORDS

Critical frequency—The highest frequency at which a vertically incident radio wave will return from the ionosphere. Above the critical frequency, radio signals pass through the **ionosphere** instead of returning to Earth.

D region—The lowest region of the ionosphere. Acts mainly to absorb energy from radio waves as they pass through it. This **ionospheric absorption** has a significant effect on signals below about 7.5 MHz during daylight.

Direct wave—A radio wave that travels in a straight line.

E region—The second lowest ionospheric region. Exists only during the day. Under certain conditions, may refract radio waves enough to return them to Earth.

F region—A combination of the two highest ionospheric regions, the F1 and F2 regions. Refracts radio waves and returns them to Earth. Its height varies greatly depending on the time of day, season of the year and amount of sunspot activity.

Ionosphere—A region of charged particles about 30 to 260 miles above the Earth that bends radio waves and returns them to Earth.

Ionospheric absorption—The D region of the ionosphere absorbs energy from radio waves that pass through it during the daytime. Limits communications range on the longer-wavelength amateur bands during the day.

Maximum usable frequency (MUF)—The highest-frequency radio signal that will reach a particular destination using E- or F-skip propagation. The MUF may vary for radio signals sent to different destinations.

Scatter—A small amount of signal that is reflected from the Earth's surface after traveling through the ionosphere. The reflected signals may go back into the ionosphere along several paths and be refracted to Earth again. Scatter can help provide communications into a station's **skip zone**.

Skip zone—An area past the maximum range of ground waves and before the range of waves returned from the ionosphere. Radio communication is difficult or impossible between stations in each other's skip zone.

Temperature inversion—A condition in the atmosphere in which a region of cool air is trapped beneath warmer air.

Troposphere—The region in Earth's atmosphere just above the Earth's surface and below the ionosphere.

Tropospheric bending—When radio waves are bent in the troposphere, they return to Earth farther away than the visible horizon.

Tropospheric ducting—A type of VHF propagation that can occur when there is a **temperature inversion**.

Visible horizon—The most distant point one can see by line of sight.

3 Radio-Wave Propagation

Radio amateurs and scientists have studied the way radio-frequency energy travels since the earliest days of wireless. We know this discipline today as the science of *propagation*. This chapter will discuss some of the factors that make radio communication possible.

High in the atmosphere, a region of thin air is electrically charged (ionized) by radiation from the sun. When ionized by solar radiation, this region, the **ionosphere**, can refract radio waves. If the wave is bent enough, it returns to Earth. As shown in Figure 3-1, if the wave is not bent enough, it travels off into space. Ham radio contacts of up to 2500 miles are possible with one *skip* off the ionosphere. Worldwide communications using several skips (or hops) can take place if conditions are right.

THE IONOSPHERE: A CLOSER LOOK

The Earth's upper atmosphere consists mainly of oxygen and nitrogen. There are traces of hydrogen, helium and several other gases. The atoms making up these gases are electrically neutral: They have no charge and exhibit no electrical force outside their own structure. The gas atoms absorb ultraviolet radiation from the sun,

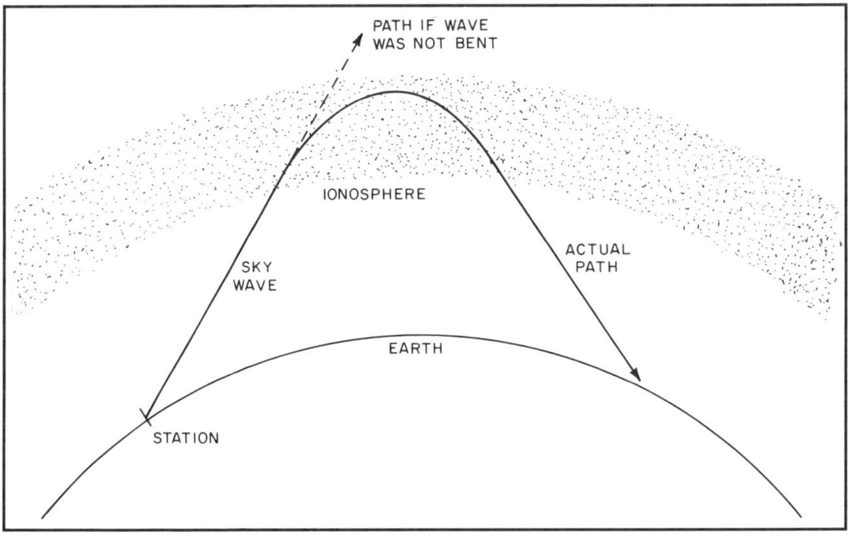

Figure 3-1—Because radio waves are bent in the ionosphere, they return to Earth far from their origin. Without refraction in the ionosphere, radio waves would pass into space.

which knocks electrons out of the atom. In this process, the atoms become positively charged. A positively charged atom is an ion. The process by which ions are formed is called *ionization*.

Several ionized regions appear at different heights in the atmosphere. Each region has a central area where the ionization is greatest. See Figure 3-2. Ionospheric regions have been given letter designations, as shown in Figure 3-3. Scientists started with the letter D just in case there were any undiscovered lower regions. None have been found, so there is no A, B or C region.

The **D region**: The lowest region of the ionosphere, the D region is in a relatively dense part of the atmosphere about 30 to 55 miles above the Earth. When the atoms in this region absorb sunlight and form ions, they don't last very long. They quickly combine with free electrons to form neutral atoms again. The amount of ionization in this region varies widely. It depends on how much sunlight hits the region. Around noon, D-region ionization is at its maximum or very close to it. By sunset, this ionization disappears. The D region is ineffective in refracting or bending high-frequency signals back to Earth.

The D region's major effect is to absorb energy from radio waves. As radio waves pass through the ionosphere, they give up energy. This sets some of the ionized particles into motion. Effects of this **ionospheric absorption** on lower frequencies (longer wavelengths) are greater than on higher frequencies. Absorption also increases proportionally with the amount of ionization. The more ionization, the more energy the radio waves lose passing through the ionosphere. Absorption is most pronounced at midday. It is responsible for the short daytime communications ranges on the lower-frequency amateur bands (160, 80 and 40 meters).

The next region of the ionosphere is the **E region**, at an altitude of about 60 to 70 miles. At this height, ionization produced by sunlight does not last very long. This

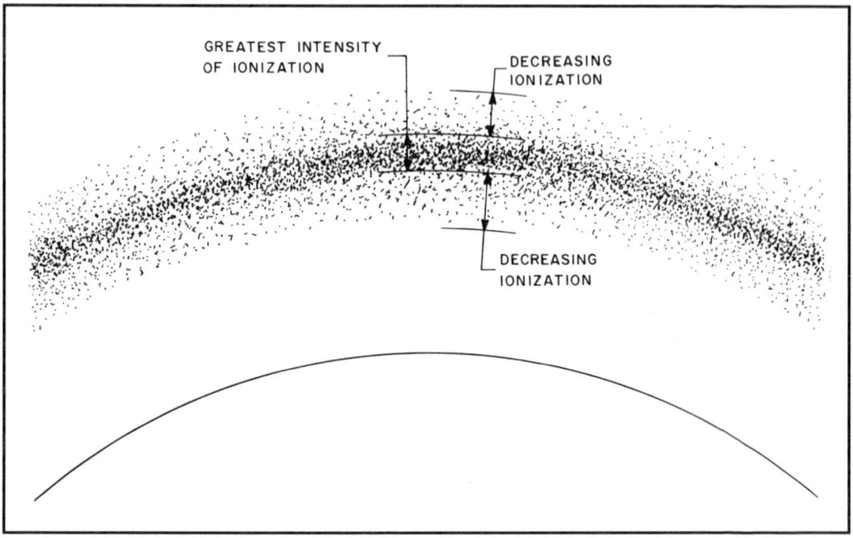

Figure 3-2—A cross-section of a layer of the ionosphere. The intensity of the ionization is greatest in the central area and decreases above and below it.

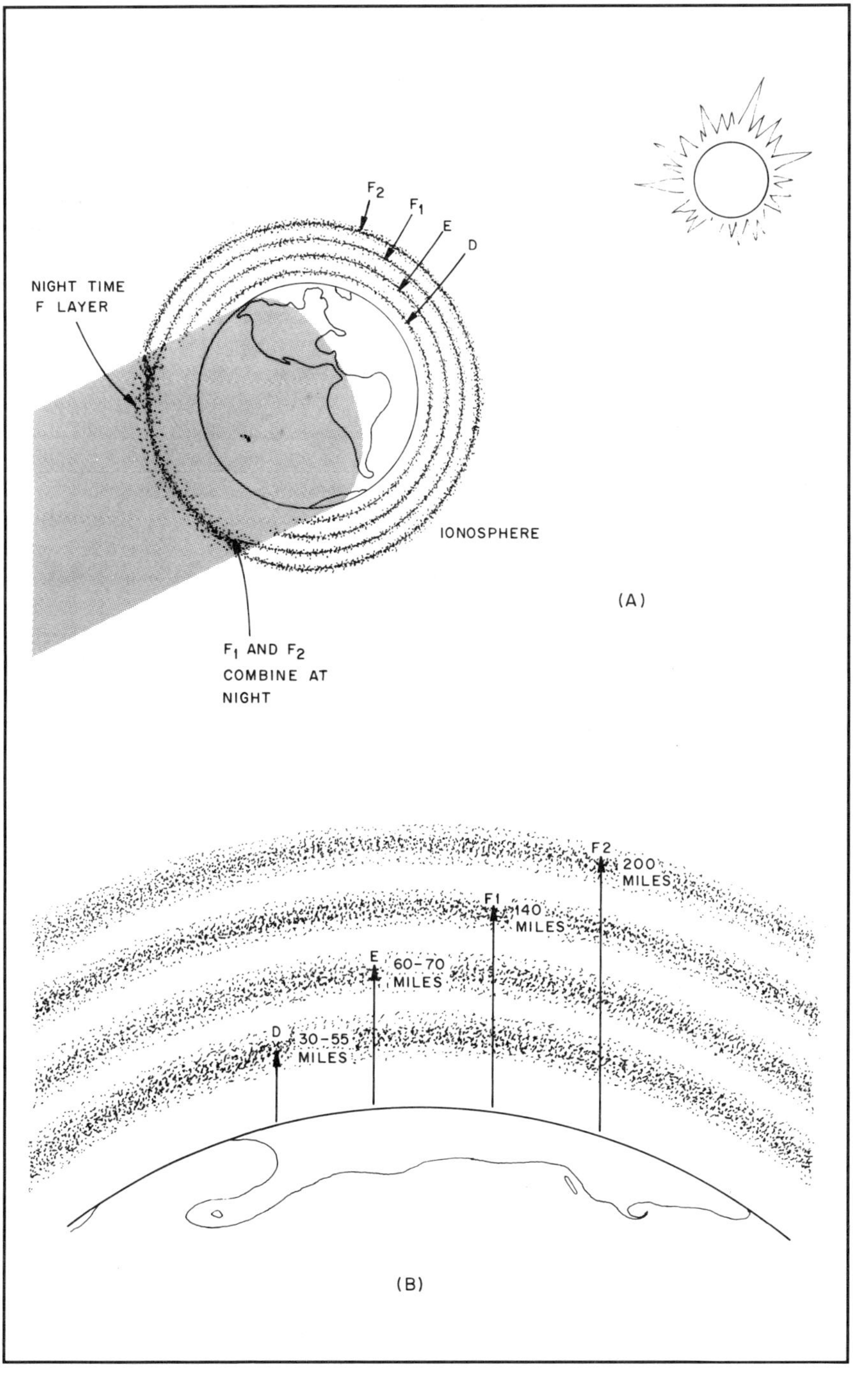

Figure 3-3—The ionosphere consists of several regions of ionized particles at different heights above Earth. At night, the D and E regions disappear and the F1 and F2 regions combine to form a single F region.

makes the E region useful for bending radio waves only when it is in sunlight. Like the D region, the E region reaches maximum ionization around midday. By early evening the ionization level is very low. The ionization level reaches a minimum just before sunrise, local time. Using the E region, a radio signal can travel a maximum distance of about 1250 miles in one hop. *Sporadic E*, or E skip, is a type of sky-wave propagation that allows long-distance communications on the VHF bands (6 meters and 2 meters—and perhaps even 222 MHz). Although, as its name implies, sporadic E occurs only during certain times of the year, it is the most common type of VHF sky-wave propagation.

The **F region**: The region of the ionosphere most responsible for long-distance amateur communication is the F region. This region is a very large region. It ranges from about 100 to 260 miles above the Earth. The height depends on season, latitude, time of day and solar activity. Ionization reaches a maximum shortly after noon local standard time. It tapers off very gradually toward sunset. At this altitude, the ions and electrons recombine very slowly. The F region remains ionized during the night, reaching a minimum just before sunrise. After sunrise, ionization increases rapidly for the first few hours. Then it increases slowly to its noontime maximum.

During the day, the F region splits into two parts, F1 and F2. The central region of the F1 region forms at an altitude of about 140 miles. For the F2 region, the central region forms at about 200 miles above the Earth. These altitudes vary with the season of the year and other factors. At noon in the summer the F2 region can reach an altitude of 300 miles. At night, the two regions recombine to form a single F region slightly below the higher altitude. The F1 region does not have much to do with long-distance communications. Its effects are similar to those caused by the E region. The F2 region is responsible for almost all long-distance communication on the amateur HF bands. A one-hop radio transmission travels a maximum of about 2500 miles using the F2 region.

[Now turn to Chapter 10 and study questions T3A01 through T3A11 and T3B01 through T3B08. Review this section as needed.]

Virtual Height

For practical purposes, think of each ionospheric region as having a definite height. The height from which a simple reflection from the region would give the same effects (observed from the ground) as the effects of the gradual bending that actually takes place is called the *virtual height*. See Figure 3-4.

The virtual height of an ionospheric region for various frequencies is determined with an ionosonde. An ionosonde is a variable-frequency transmitter and receiver. It directs radio energy vertically and measures the time required for the signal to make a round-trip. As the frequency increases, there is a point where no energy will return. The highest frequency at which vertically incident waves return to Earth is the **critical frequency**. Signals at higher frequencies pass through the ionosphere into space.

Radiation Angle and Skip Distance

A radio wave at a low angle above the horizon requires less refraction to bring the wave back to Earth. This is why we want antennas with low radiation angles to work DX.

Figure 3-5 illustrates some of the effects of radiation angle. The high-angle sky

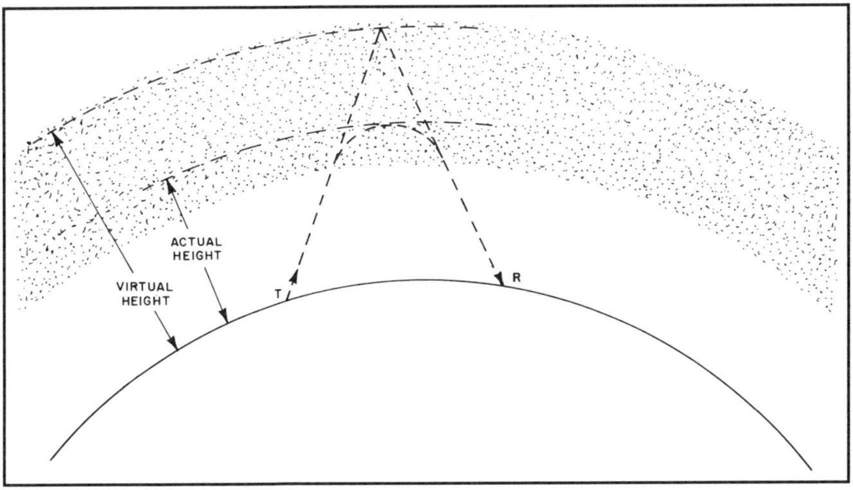

Figure 3-4—The virtual height of a region in the ionosphere is the height at which a simple reflection would return the wave to the same point as the gradual bending that actually takes place.

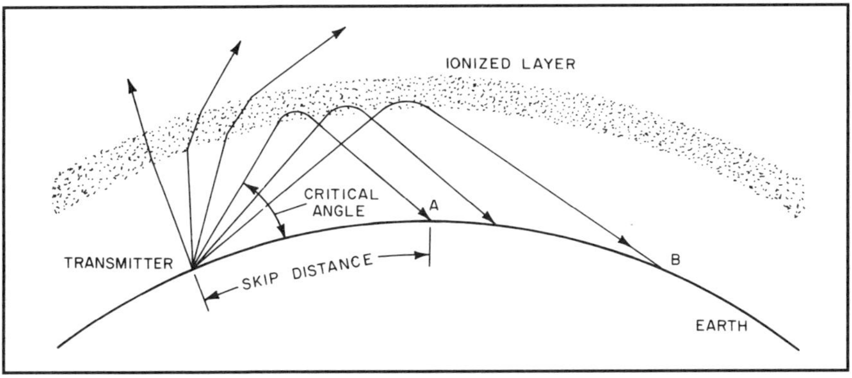

Figure 3-5—Ionospheric propagation. The waves at left leave the transmitter above the critical angle—they are refracted in the ionosphere, but not enough to return to Earth. The wave at the critical angle will return to Earth. The lowest-angle wave will return to Earth farther away than the wave at the critical angle. This explains the emphasis on low radiation angles for DX work.

waves are bent only slightly in the ionosphere and pass through it. The wave at the somewhat lower angle is just able to return. In daylight it might be returned from the E region. The point of return for high angle waves is relatively close to the transmitting station. The lowest angle wave returns farther away, to point B.

The highest radiation angle that will return a radio wave to Earth under specific ionospheric conditions is the critical angle. Waves meeting the ionosphere at greater than the critical angle will pass through the ionosphere into space.

Maximum Usable Frequency

There is little doubt that the critical frequency is important to amateur communication. More interesting to radio amateurs is the frequency range over which communication can be carried via the ionosphere. Most amateurs want to know the **maximum usable frequency** (abbreviated MUF). The MUF applies to a particular direction and distance at the time of day when you want to communicate. The MUF is the highest frequency that allows a radio wave to reach the desired destination using E- or F-region propagation. The MUF is subject to seasonal variations as well as changes throughout the day, based on the amount of ultraviolet solar radiation.

Why is the MUF so important? If we know the MUF, we can make accurate predictions about which bands give the best chance for communication over a particular path. Predictions concerning amateur bands affording the best chance for communication via particular paths can be made. For example, we wish to contact a station in Australia. We know that the MUF for the path between our station and the one in Australia is 17 MHz at the time we wish to make our contact. The closest amateur band below that frequency is the 20-meter, or 14-MHz, band. The 20-meter band offers our best chance to contact that location, then. There is no single MUF for a given location at any one time. The MUF will vary depending on the direction and distance to the station we wish to contact. Understanding and using the MUF is one way you can change your study of propagation from guesswork into a science.

[Now turn to Chapter 10 and study exam questions T3B09 through T3B11. Review this material as needed.]

Solar Activity

It should be clear that the sun plays the major role in the ionization of different regions in the ionosphere. The sun is the dominant factor in sky-wave communication. To communicate effectively over long distances, you must understand how solar conditions will affect your radio signals.

You know about the day-to-day sun-related cycles on Earth, such as the time of day and the season of the year. Conditions affecting radio communication vary with these same cycles. Long-term and short-term solar cycles also influence propagation in ways that are not so obvious. The condition of the sun at any given moment has a very large effect on long-distance radio communication. The sun is what makes propagation prediction an inexact science.

Man's interest in the sun is older than recorded history. Sunspots are dark regions that appear on the surface of the sun. They were observed and described thousands of years ago. Observers noted that the number of sunspots increased and decreased in cycles. The solar observatory in Zurich, Switzerland, has been recording solar data on a regular basis since 1749. The solar cycle that began in 1755 was designated Cycle 1. The low sunspot numbers marking the end of Cycle 21 and the beginning of Cycle 22 occurred in September of 1986. Sunspot cycles also vary a great deal. For example, Cycle 19 peaked with a smoothed mean sunspot number of over 200. Cycle 14 peaked at only 60.

Maximum ionization occurs during a sunspot cycle peak. High sunspot numbers usually mean good worldwide radio communications. During sunspot cycle peaks, the 20-meter amateur band is open to distant parts of the world almost continuously. The Cycle 19 peak in 1957 and 1958 was responsible for the best propagation conditions in the history of radio. Sunspot cycles do not follow consistent patterns.

There can be periods of high activity that seem to come from nowhere during periods of low sunspot activity.

There is an important clue to anticipating variations in solar radiation levels. Radio propagation changes resulting from these variations can be predicted. This clue is the time it takes the sun to rotate on its axis, approximately 27 days. Active areas capable of influencing propagation can recur at four-week intervals. These active areas may last for four or five solar rotations. High MUF and good propagation for several days indicates similar conditions may develop approximately 27 days later.

Another useful indication of solar activity is solar flux, or radio energy coming from the sun. Increased solar activity produces higher levels of solar energy. More solar energy produces greater ionization in the ionosphere. Scientists use sophisticated receiving equipment and large antennas pointed at the sun to measure solar flux. A number called the *solar-flux index* is given to represent the amount of solar flux. The solar-flux index is gradually replacing the sunspot number as a means of predicting radio-wave propagation.

The solar-flux measurement is taken at 1700 UTC daily in Ottawa, Canada, on 2800 MHz (10.7 centimeters). The information is then transmitted by the National Institute of Science and Technology station WWV in Fort Collins, Colorado. Both the sunspot number and the solar-flux measurements tell us similar things about solar activity. To get the sunspot number, the sun must be visible. The solar-flux measurement may be taken under any weather conditions.

Using the Solar Flux

We can use the solar-flux numbers to make general daily predictions of band conditions. The solar flux varies directly with the activity on the sun. Values range from around 60 to 250 or so. Flux values in the 60s and 70s mean fair to poor propagation conditions on the 14-MHz (20-meter) band and higher frequencies. Values from 90 to 110 or so indicate good conditions up to about 24 MHz (12 meters). Solar-flux values over 120 indicate very good conditions on 28 MHz (10 meters) and even up to 50 MHz (6 meters) at times. On 6 meters, the flux values must be in the 200s for reliable long-range communications. The higher the frequency, the higher the flux value required for good propagation.

Sky-wave or *skip propagation* is primarily an HF propagation mode; it is rare at VHF. When the solar-flux values are very high, however, you may find enhanced propagation on 6 meters and even on 2 meters sometimes. If you notice strong signals on 10 meters from stations 500 to 600 miles away, this is a good indication you should look for skip on 6 and 2 meters.

The Scatter Modes

All electromagnetic-wave propagation is subject to scattering influences. These alter idealized patterns to a great degree. The Earth's atmosphere, ionospheric regions and any objects in the path of radio signals scatter the energy. Understanding how **scatter** propagation takes place helps us use this mode to our advantage.

Forward Scatter

There is an area between the outer limit of ground-wave propagation and the point where the first signals return from the ionosphere. This area, called the **skip zone**, is often described as if communications between stations in each other's skip zone were impossible. Actually, some of the transmitted signal is scattered in the atmosphere, so the signal can be heard over much of the skip zone. See Figure 3-6. You usually need a very sensitive receiver and good operating techniques to hear these signals, as they are usually weak and distorted.

The troposphere is a region of the atmosphere below the ionosphere. VHF "tropo" scatter is usable out to about 500 miles from the transmitting station.

Ionospheric scatter, mostly from the height of the E region, is most apparent at frequencies up to about 60 or 70 MHz. This type of forward scatter may be usable out to about 1200 miles. Ionospheric scatter propagation is most noticeable on frequencies above the MUF.

Another means of ionospheric scatter is provided by meteors entering Earth's atmosphere. As the meteor passes through the ionosphere, a trail of ionized particles forms. These particles can scatter radio energy. This ionization is short-lived and can show up as short bursts of signals with little communication value. There may be

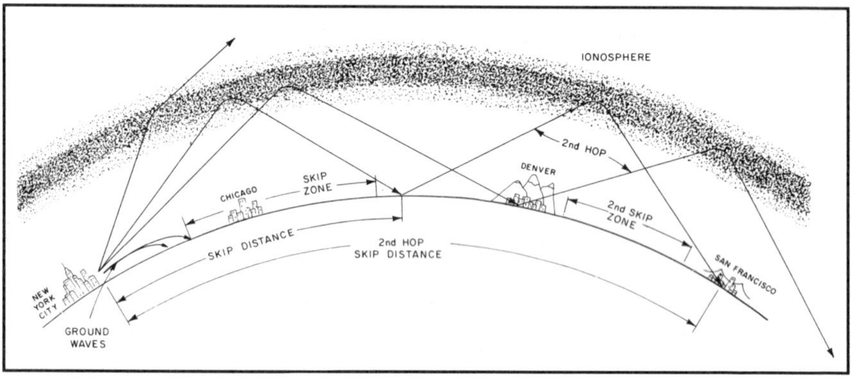

Figure 3-6—There is an area between the farthest reaches of ground-wave propagation and the closest return of sky waves from the ionosphere. This region is known as the skip zone.

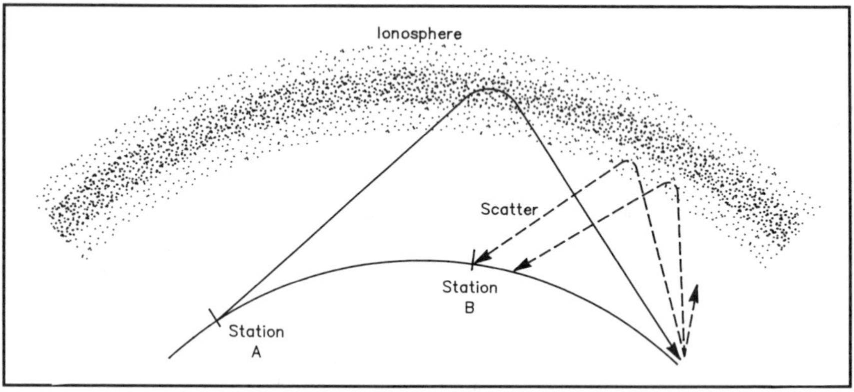

Figure 3-7—When radio waves strike the ground after passing through the ionosphere, some of the signal may reflect back into the ionosphere. If some of the signal is reflected from the ionosphere toward the transmitting station, the energy may be scattered back into the skip zone.

longer periods of usable signal levels, lasting up to a minute or more. Meteor scatter is most common between midnight and dawn. It peaks between 5 and 7 AM local time. Meteor scatter is an interesting method of amateur communication at 21 MHz or higher, especially during periods of low solar activity.

Scatter

You can observe a complex form of scatter when you are working very near the MUF. The transmitted wave is refracted back to Earth at some distant point. This may be an ocean area or land mass. A portion of the transmitted signal reflects back into the ionosphere. Some of this signal comes toward the transmitting station. The reflected wave helps fill in the skip zone, as shown in Figure 3-7.

Scatter signals are generally weak and subject to distortion, because the signal may arrive at the receiver from many different directions. Scatter (sometimes called *backscatter*) is usable from just beyond the local range out to several hundred miles. Under ideal conditions, scatter is possible over 3000 miles or more. The term "sidescatter" is more descriptive of what probably happens on such long paths, however.

[Now study exam questions T3C01, T3C02 and T3C11 in Chapter 10. Review the material in this section as needed.]

LINE-OF-SIGHT PROPAGATION

In the VHF and UHF range (above 144 MHz), most propagation is by space wave. Figure 3-8 shows how space waves travel within the range of the visible horizon. Space waves are made up of a **direct wave**, and usually one or more reflected waves. The direct wave and reflected wave arrive at the receiving station slightly out of phase. They may cancel or reinforce each other to varying degrees. This depends on the actual path of the reflected wave(s). Effects are most pronounced when either, or both, stations are mobile. When one or both stations are moving, the paths are constantly changing. The waves may alternately reinforce and cancel each other. This causes a rapid fluttering sound (called "picket fencing").

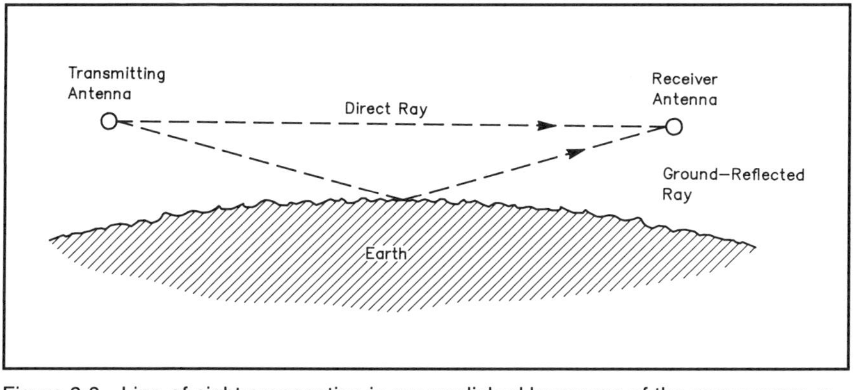

Figure 3-8—Line-of-sight propagation is accomplished by means of the space wave, a combination of a direct ray and one or more reflected rays.

Tropospheric Bending and Ducting

The **troposphere** is an atmospheric region close to the Earth's surface. **Tropospheric bending** is evident over a wide range of frequencies. It is most useful in extending the range of stations operating at VHF and UHF, especially at 144 MHz and above. (There is very little sky-wave propagation on VHF and higher frequencies.) Instead of gradual changes in air temperature, pressure and humidity, distinct regions may form in the troposphere. Adjacent regions having significantly different densities will bend radio waves passing between regions. (In the same way, light is bent when it passes from air into water. That's why a spoon in a glass of water seems to bend when it enters the water.)

The **visible horizon** is the most distant point one can see. It is limited by the height of the observer above ground. Slight bending of radio waves occurs in the troposphere. This will cause signals to return to Earth somewhat beyond the geometric horizon. See Figure 3-9.

Under normal conditions, the temperature of the air gradually decreases with increasing height above ground. When there's a stable high-pressure system, a mass of warm air may overrun cold air. When warm air covers cold air, we have a

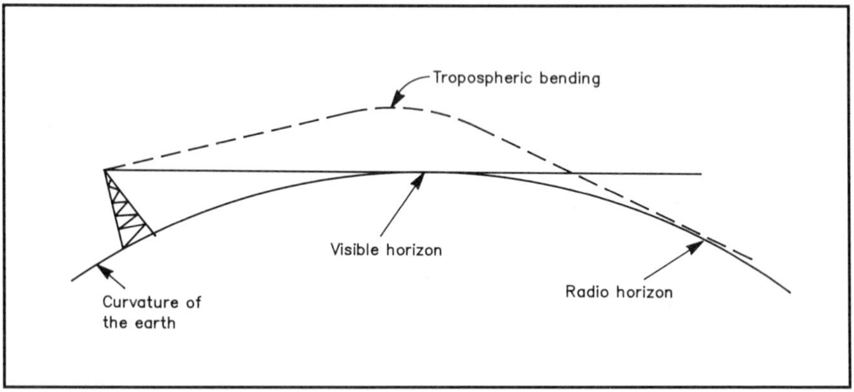

Figure 3-9—Under normal conditions, tropospheric bending causes radio waves to be returned to Earth beyond the visible horizon.

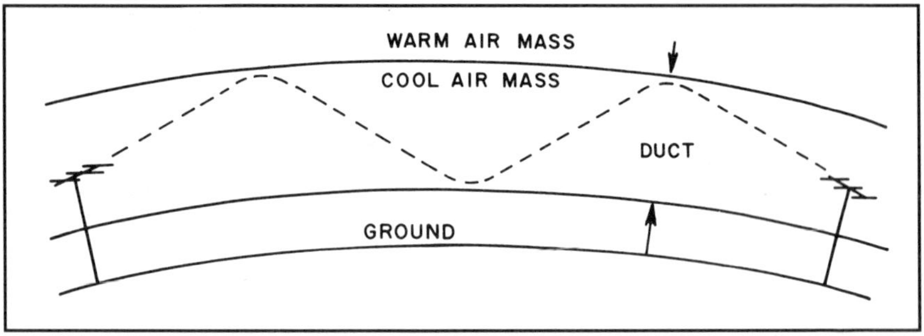

Figure 3-10—When a cool air mass is overrun by a mass of warmer air, a duct is formed, allowing VHF radio signals to travel great distances with little attenuation.

temperature inversion. Radio waves can be trapped below the warm air mass. They can travel great distances with little loss. The area between the Earth and the warm air mass is known as a *duct*. See Figure 3-10.

The term for radio-wave propagation through a duct is **tropospheric ducting**. As with tropospheric bending, we find ducting primarily at VHF frequencies, 144 MHz and higher. Ducts usually form over water, though they can form over land as well. The lowest usable frequency for duct propagation depends on two factors: the depth of the duct and the amount the refractive index changes at the air-mass boundary.

[At this point, turn to Chapter 10 and study questions T3C03 through T3C10. Review any material necessary.]

KEY WORDS

Ammeter—A test instrument that measures current.

ANSI RF protection guide—Standards formulated by the American National Standards Institute that provide RF exposure guidelines.

Crystal calibrator—(see **marker generator**).

Directional wattmeter—(see **wattmeter**)

Dummy antenna (dummy load)—A station accessory that allows you to test or adjust transmitting equipment without sending a signal out over the air.

Marker generator—A high-stability oscillator that produces reference signals at known frequency intervals. It can be used to calibrate receiver and transmitter tuning dials. Also called a **crystal calibrator**.

Multimeter—An electronic test instrument used to measure current, voltage and resistance in a circuit. Describes all meters capable of making these measurements, such as the *volt-ohm-milliammeter (VOM)*, *vacuum-tube voltmeter (VTVM)* and *field-effect transistor VOM (FET VOM)*.

National Electrical Code—A set of guidelines governing electrical safety, including antennas.

Radio Frequency Protection Guide of the American National Standards Institute—Standards that recommend RF exposure limits for humans.

Reflectometer—A test instrument used to indicate standing wave ratio (SWR) by measuring the forward power (power from the transmitter) and reflected power (power returned from the antenna system).

S meter—A meter in a receiver that shows the relative strength of a received signal.

Signal generator—A test instrument that produces a stable low-level radio-frequency signal. The signal can be set to a specific frequency and used to troubleshoot RF equipment.

SWR meter—(see **reflectometer**)

Voltmeter—A test instrument used to measure voltage.

Wattmeter—Also called a power meter, a test instrument used to measure the power output (in watts) of a transmitter. A directional wattmeter measures both forward and reflected power.

WWV and **WWVH**—Radio stations that broadcast the precise time on precise frequencies. Hams tune in to these stations to calibrate transceiver dials (and to set their clocks).

4 Amateur Radio Practice

It's easy to form good operating habits in Amateur Radio. You will need to understand the fundamentals of radio. Then you can maintain your station to operate within FCC rules and regulations. Also, if you're like most amateurs, you'll want to put out the best possible signal. Know the principles behind the operation of your station. Correctly adjust your equipment and measure its performance. Take advantage of available test equipment. You should have no problem keeping your station "up to par."

ELECTRICAL WIRING SAFETY

Your station equipment makes use of ac line voltage. This voltage can be dangerous. The equipment also generates additional potentially lethal voltages of its own. You should be familiar with some basic precautions. Your own safety and that of others depends on it.

Power-Line Connections

In most residential systems, three wires come in from outside to the distribution board. Older systems may use only two wires. In the three-wire system, the voltage between the two "hot" wires is normally 240. The third wire is neutral and is grounded. Half of the total voltage appears between each of the hot wires and neutral, as shown in Figure 4-1. Lights, appliances and 120-V outlets are divided as evenly as possible between the two sides of this circuit. Half of the load connects between one hot wire and the neutral. The other half of the load connects between the other hot wire and neutral.

Heavy appliances, such as electric stoves and most high-power amateur amplifiers, are designed for 240-V operation. Connect these across the two hot wires. Both ungrounded wires should be fused. The neutral wire should never have a fuse or switch in it.

High-power amateur amplifiers use the full 240 V. It only takes half as much

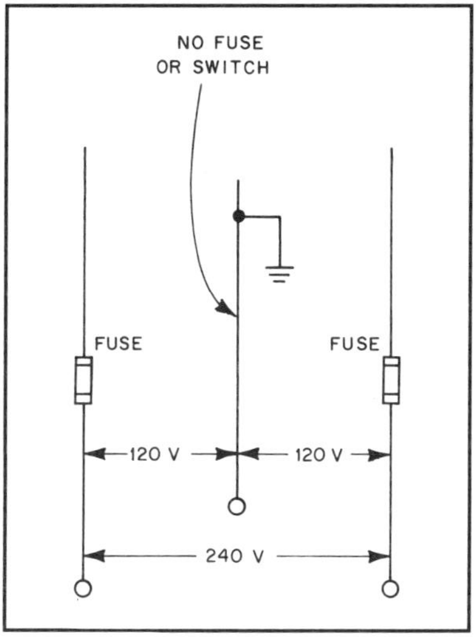

Figure 4-1—In home electrical systems, two wires carrying 240 volts are split between the house circuits carrying 120 volts each. Some heavy appliances use the full 240 V. As shown, a fuse should be placed in each hot wire, but no fuse or switch should be placed in the neutral wire.

current to supply the same power as with a 120-V line. The power supply circuit efficiency improves with the higher voltage and lower current. A separate 240-V circuit for the amplifier ensures that you stay within the capacity of the 120-V circuits in your shack. An amplifier drawing more current than the house wiring can handle could cause the house lights to dim. This is because the heavy current causes the power-line voltage to drop.

Three-Wire 120-V Power Cords

State and national electrical-safety codes require three-wire power cords on many 120-V tools and appliances. Two of the conductors (the "hot" and "neutral" wires) power the device. The third conductor (the safety ground wire) connects to the metal frame of the device. See Figure 4-2. The "hot" wire is usually black or red. The "neutral" wire is white. The frame/ground wire is green or sometimes bare.

Let's look at the power cord connections for a transmitter power supply. The power supply will have a transformer in it. When you attach the power cord, the black wire will attach to one end of the fuse. The white wire will go to the side of the transformer primary winding without a fuse. The green (or bare) wire will attach to the chassis.

When plugged into a properly wired mating receptacle, a three-contact plug connects the third conductor to an earth ground. This grounds the appliance chassis or frame and prevents the possibility of electric shock. A defective power cord that shorts to the case of the appliance will simply blow a fuse. Without the ground connection, the case could carry the full line voltage, presenting a severe shock hazard. All commercially manufactured electronic test equipment and most ac-operated amateur equipment uses these three-wire cords. Adapters are available for use where older electrical installations do not have mating receptacles. The lug from the adapter

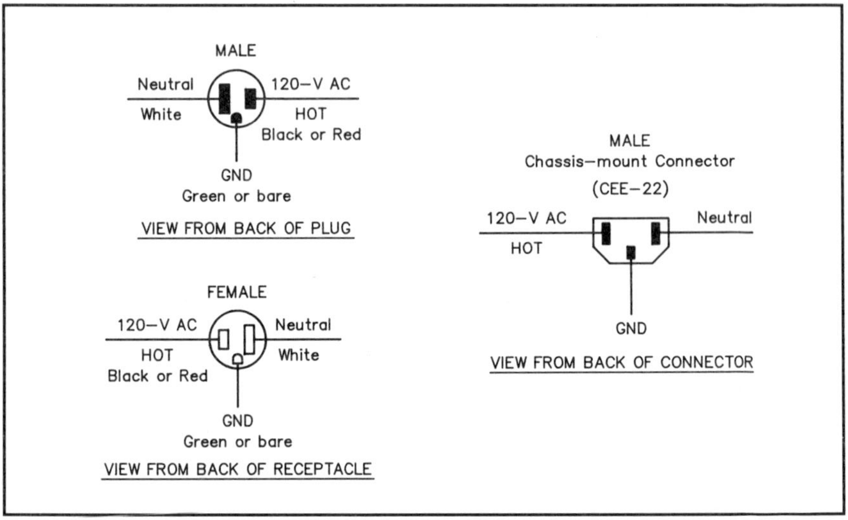

Figure 4-2—Correct wiring technique for 120-V power cords and receptacles. The white wire is neutral, the green wire is ground and the black or red wire is the hot lead. Notice that the devices are shown as viewed from the back, or wiring side.

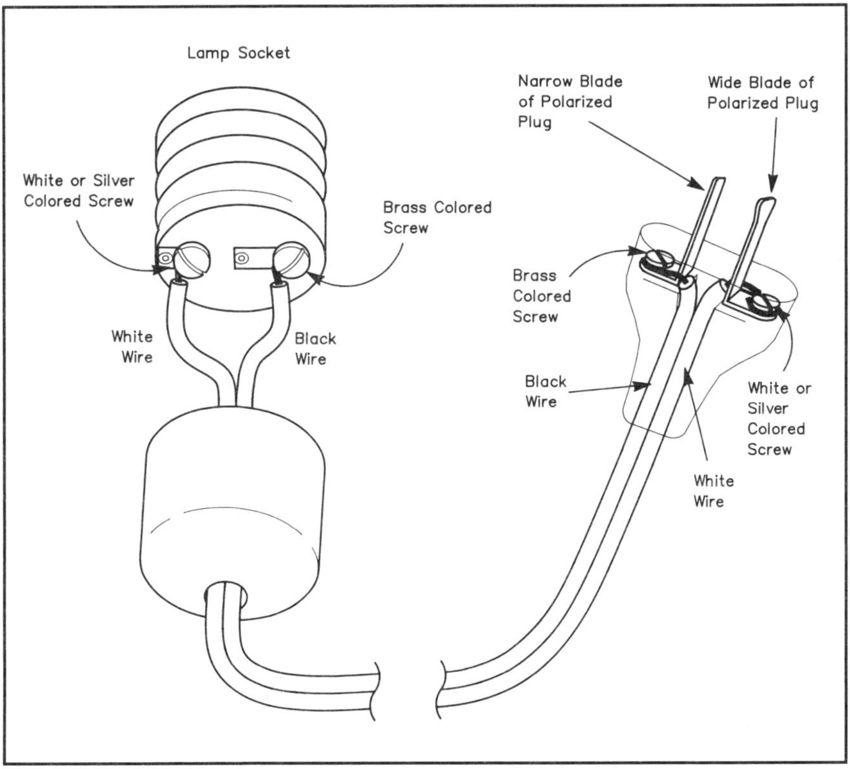

Figure 4-3—The correct way to wire a lamp socket. Connect the white wire to the white or silver terminal and the black wire to the brass-colored terminal. Connect the wires to a polarized plug the same way. This ensures that the lamp-socket shell connects to the neutral side of the ac supply.

must be attached under the cover-plate screw. The outlet (and outlet box) must be grounded for this to be effective. Power wires coming into the electric box inside a flexible metal covering provide grounding through the metal covering. The common name for this type of wire is armored cable.

A "polarized" two-wire plug has one blade that is wider than the other. The mating receptacle will accept the plug only one way. This ensures that the hot and neutral wires in the appliance connect to the appropriate wires in the house electrical system. Consider what happens without this polarized plug and receptacle. The power switch in the equipment will be in the hot wire when the plug is inserted one way. It will be in the neutral wire when inserted the other way. This can present a dangerous condition. It is possible for the equipment to be "hot" even with the switch off. With the switch in the neutral line, the hot line may be connected to the equipment chassis. An unsuspecting operator could form a path to ground by touching the case, and might be electrocuted or receive a nasty shock!

Wiring an outlet or lamp socket properly is important. See Figure 4-3. The black (hot) wire should be connected to the brass terminal on the lamp socket or outlet. The white (neutral) wire should be connected to the white or silver-colored terminal. This

Table 4-1

Current-Carrying Capability of Some Common Wire Sizes

Copper Wire Size (AWG)	*Allowable Ampacity (A)*	*Max Fuse or Circuit Breaker Size (A)*	*Permissible Load (A)*[2]
6	55	50	40
8	40	40	32
10	30	30	24
12	25 (20)[1]	20	16
14	20(15)[1]	15	12

[1]The National Electrical Code limits that fuse or circuit breaker size (and, as such, the maximum allowable circuit load) to 15 A for #14 AWG and to 20 A for #12 AWG conductors.

[2]Adjustments from these values may be possible. Consult your local electrical inspector, a licensed electrician or a registered electrical engineer for assistance.

will ensure that the proper blade of the plug connects to the hot wire (that the polarity is correct). This practice is especially important when wiring lamp sockets. The brass screw of the socket connects to the center pin in the socket. With this pin as the "hot" connection, there is much less shock hazard. Anyone unscrewing a bulb from a correctly wired socket has to reach inside the socket to get a shock. In an incorrectly wired socket, the screw threads of the bulb are "hot." A dangerous shock may result when replacing a bulb.

Current Capacity

There is another factor to take into account when you are wiring an electrical circuit. It is the current-handling capability (ampacity) of the wire. Table 4-1 shows the current-handling capability of some common in-wall residential wire sizes. The table shows that number 14 wire could be used for a circuit carrying 15 A. You must use number 12 (or larger) for a circuit carrying 20 A. Wires smaller than number 14 may not be used for in-wall residential wiring.

To remain safe, don't overload the ac circuits in your home. The circuit breaker or fuse rating is the maximum load for the line at any one time. Simple addition can be used to calculate the current drawn on the circuit serving your ham radio equipment. Most equipment has the power requirements printed on the back. If not, the owner's manual should contain such information. Calculations must include any other household equipment or plugged-in appliances on the same line (lighting, air-conditioning, fans, and so on). The sum of all the power requirements connected to the same circuit must not exceed the permissible loads shown in Table 4-1.

Do not put a larger fuse in an existing circuit—too much current could be drawn (see Table 4-1 for the maximum fuse size for the existing wire size). The wires would become hot and a fire could result. Installing too large of a fuse or circuit breaker violates the **National Electrical Code** and most local ordinances. Such violations may also invalidate your fire insurance! The National Electrical Code is used across the US as the basis for electrical safety relating to power wiring and antennas.

One final precaution: It's important that all family members know the location of the main electrical box in your home, and how to turn it off. In case of electrical emergency, this precaution could save a life.

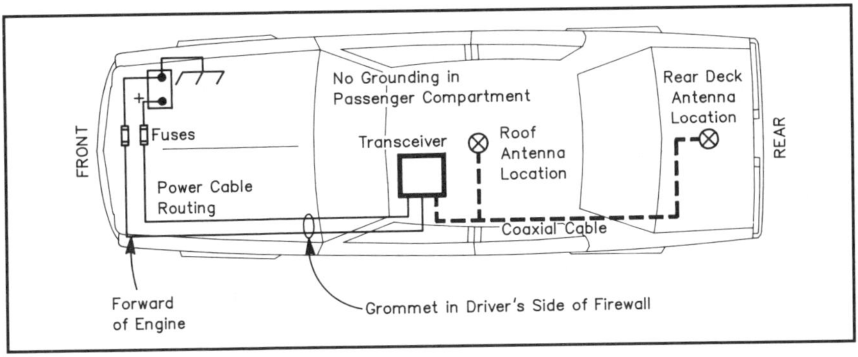

Figure 4-4—A wiring diagram for a typical mobile transceiver installation.

Mobile Power Connections

Installing a mobile transceiver requires some thought. It's best to use a direct connection to the vehicle battery to power a radio from a car. This minimizes voltage drops and electrical noise. Although one side of the battery is connected to the car frame, you should run positive and negative wires, and install fuses in series with both the red and black leads. The fuses should be located at the battery. See Figure 4-4.

You'll want to prevent unauthorized persons from using your mobile radio once it's installed. You can do that easily by disconnecting and removing the microphone when you leave the vehicle.

[Now turn to Chapter 10 and study questions T4A01 through T4A07. Review this section if you have any difficulty.]

Electrical Equipment Safety

All dangerous voltages in equipment should be made inaccessible. A good way to ensure this is to enclose all equipment in metal cabinets. That way no "hot" spots can be reached. Don't forget any component shafts that might protrude through the front panel. If a control shaft is hot, protect yourself from accidental contact by using an insulated shaft extension or insulated knob. Another precaution to take is to avoid wearing loose jewelry when working around potentially dangerous pieces of equipment, such as an amplifier.

What is a Good Ground?

Since station grounding is a good idea for several reasons, it's worth discussing in detail. Each metal enclosure must be effectively grounded. This is generally accomplished by connections to the equipment grounding conductor (green or bare wire) contained in the power cord or premises wiring system. If a connection is made to a separate ground rod for radio frequency (RF) purposes, that ground rod must be bonded to the building grounding-electrode system. Then, if a failure occurs inside a piece of equipment, the metal case will never present a shock hazard. The fuse will blow or the circuit breaker will trip instead.

Most amateurs connect their equipment to a ground rod driven into the ground as close to the shack as possible. This provides a ground connection for stray RF signals. For a few dollars you can purchase a ½- or ⅝-inch-diameter, 8-foot-long

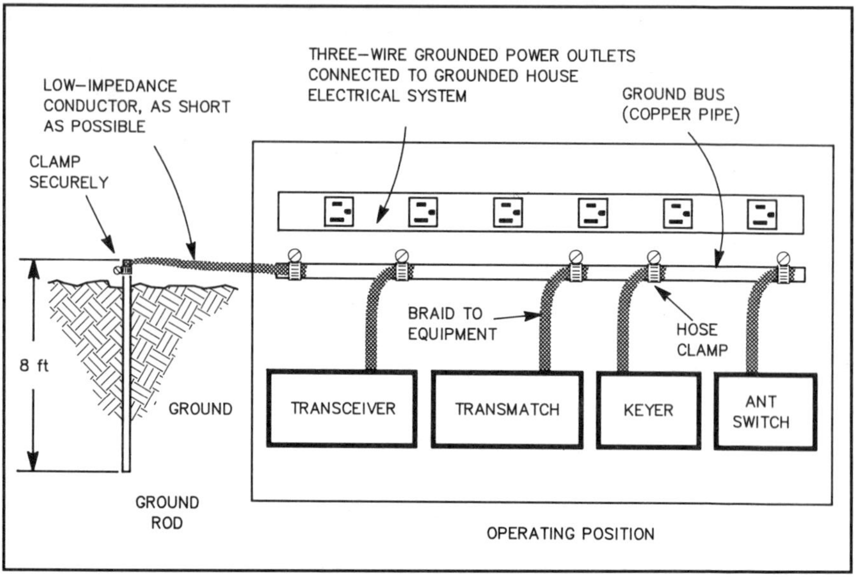

Figure 4-5—An effective station ground looks like this: all equipment is bonded together with a strong conductor such as copper flashing or coaxial cable braid. It then ties into a good earth ground—an 8-foot ground rod located as close to the station as possible.

ground rod at any electrical supply store. (Eight feet is the shortest practical length for a station ground rod.) Drive this copper-clad steel rod into the ground outside of your house, as close to your station location as possible. (Don't settle for the short, thin "ground rods" sold by some discount electronics stores. The copper cladding on the outside of most of these steel rods is very thin. They begin to rust almost immediately when they are put in the ground.)

Run a heavy copper wire (number 8 or larger) from your shack and attach it to the rod with a clamp. You can purchase the clamp when you buy the ground rod. Heavy copper strap or flashing (sold at hardware or roofing-supply stores) is even better. Figure 4-5 shows one method of grounding each piece of equipment in your station. It's important to connect the chassis of each piece of station equipment to an effective ground connection. Keep the cable between your station and the earth ground as short as possible.

The National Electrical Code requires all ground rods to be connected to form a single grounding system. These connections must use number 8 or larger copper wire. Have a qualified electrician perform this work if you are not familiar with the National Electrical Code and common wiring practices.

POWER-SUPPLY SAFETY

Safety must always receive careful consideration during the design and construction of any power supply. Power supplies can produce potentially lethal currents and voltages. Be careful to guard against accidental exposure to these currents and voltages. Use electrical tape, insulated tubing (spaghetti) or heat-shrink tubing to cover exposed wires. This includes component leads, component solder terminals and tie-down points. Whenever possible, connectors used to mate the power supply to the outside world should be of an insulated type. They should be designed to prevent

Table 4-2

Effects of Electric Current Through the Body of an Average Person

Current (1 Second Contact)	*Effect*
1 mA	Just perceptible.
5 mA	Maximum harmless current.
10-20 mA	Lower limit for sustained muscular contractions.
30-50 mA	Pain.
50 mA	Pain, possible fainting. "Can't let go" current.
100-300 mA	Normal heart rhythm disrupted. Electrocution if sustained current.
6 A	Sustained heart contraction. Burns if current density is high.

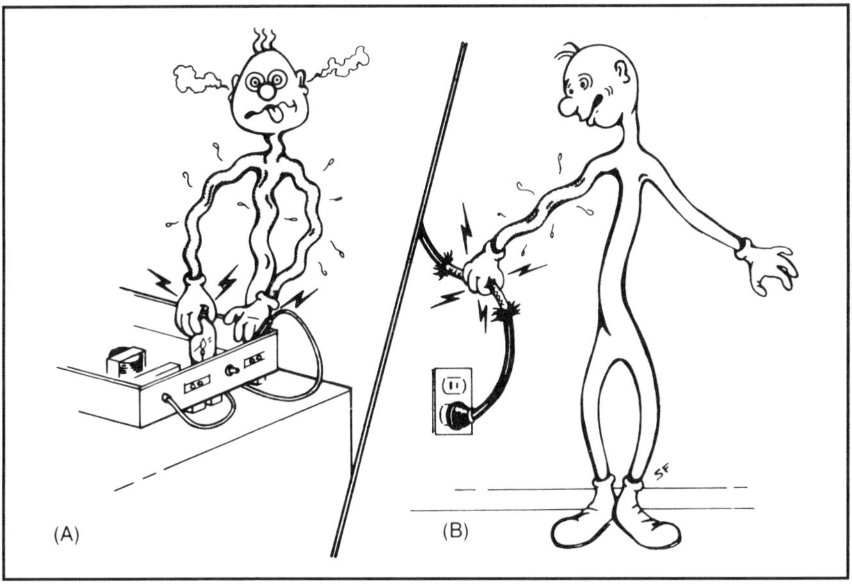

Figure 4-6—The path from the electrical source to ground affects how severe an electrical shock will be. The most dangerous path (from hand to hand directly through the heart) is shown at A. The path from one finger to the other shown at B is not quite so dangerous.

accidental contact with the voltages present. AC power to the supply should be controlled by a clearly labeled front-panel switch. That way it can be seen and reached easily in an emergency.

You should never underestimate the potential hazard when working with electricity. Table 4-2 shows some of the effects of electric current—as little as 100 mA, or 1/10 A, can be fatal! As the saying goes, "It's volts that jolts, but it's mills that kills." Low-voltage power supplies may seem safe, but even battery-powered equipment should be treated with respect. The minimum voltage considered dangerous to humans is 30 volts. These voltage and current ratings are only general guidelines. Automobile batteries are designed to provide very high current (as much as 200 A) for short periods when starting a car. This much current can kill you, even at 12 volts. You will feel pain if the shock current is in the range of 30 to 50 mA. [Question T4A10 indicates a current as small as 2 mA (1/500 A) can be painful.]

A few factors affect just how little voltage and current can be considered dangerous. One factor is skin resistance. The lower the resistance of the path, the more current that will pass through it. If you perspire heavily, you may get quite a bit more severe shock than if your skin was dry. Another factor is the path through the body to ground. As Figure 4-6 shows, the most dangerous path is from one hand to the other. This path passes directly through a person's heart. Even a very minimal current can cause heart failure and death. Current passing from one finger to another on the same hand will not have quite such a serious effect. For this reason, if you must troubleshoot a live circuit, keep one hand behind your back or in your pocket. If you do slip, the shock may not be as severe as if you were using both hands.

If you should discover someone who is being burned by high voltage, immediately turn off the power, call for help and give cardio-pulmonary resuscitation (CPR). Be sure all family members know how to turn off power at the main power switch. These measures could save the life of a friend or family member.

[Now turn to Chapter 10 and study exam questions T4A08 through T4A15. Review any topics necessary before proceeding.]

USING TEST EQUIPMENT

The Voltmeter

The **voltmeter** is an instrument used to measure voltage. It is a basic meter movement with a resistor in series, as shown in Figure 4-7. The current multiplied by

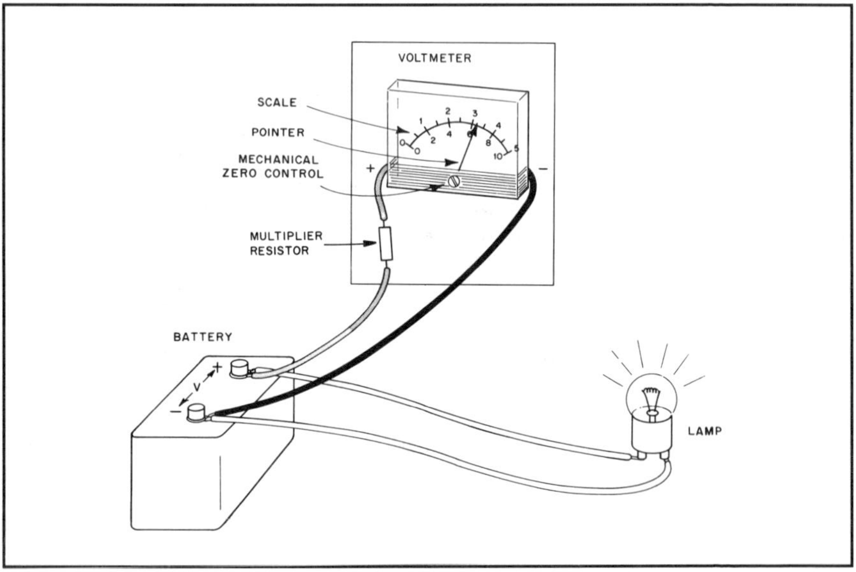

Figure 4-7—When you use a voltmeter to measure voltage, the meter must be connected in parallel with the voltage you want to measure.

the resistance will be the voltage drop across the resistance. An instrument used this way is calibrated in terms of the voltage drop across the resistor to read voltage.

A high-value resistor provides a high impedance input for the meter. An ideal voltmeter would have an infinite input impedance. It would not affect the circuit under test in any way. Real-world voltmeters, however, have a finite value of input impedance. There is a possibility that the voltage you are measuring will change. This is because the meter adds some load to the circuit when you connect the voltmeter.

Use a voltmeter with a very high input impedance compared with the impedance of the circuit you are measuring. This prevents the voltmeter from drawing too much current from the circuit. Excessive current drawn from the circuit would significantly affect circuit operation. The input impedance of an ordinary voltmeter is about 20 kilohms per volt. Voltage measurements are made by placing the meter in parallel with the circuit voltage to be measured.

The range of a meter can be extended to measure a wide range of voltages. This is accomplished by adding resistance in series with it. Changing the multiplier resistor in Figure 4-7 changes the measuring range of the meter.

The Ammeter

The **ammeter** depends on a current flowing through it to deflect the needle. The ammeter is placed in series with the circuit. That way, all the current flowing in the circuit must pass through the meter. Many times the meter cannot handle all the current. The range of the meter can be extended by placing resistors in parallel with the meter to provide a path for part of the current. A shunt resistor is shown in Figure 4-8. Selecting different shunt resistors changes the current-measuring range. The shunt-

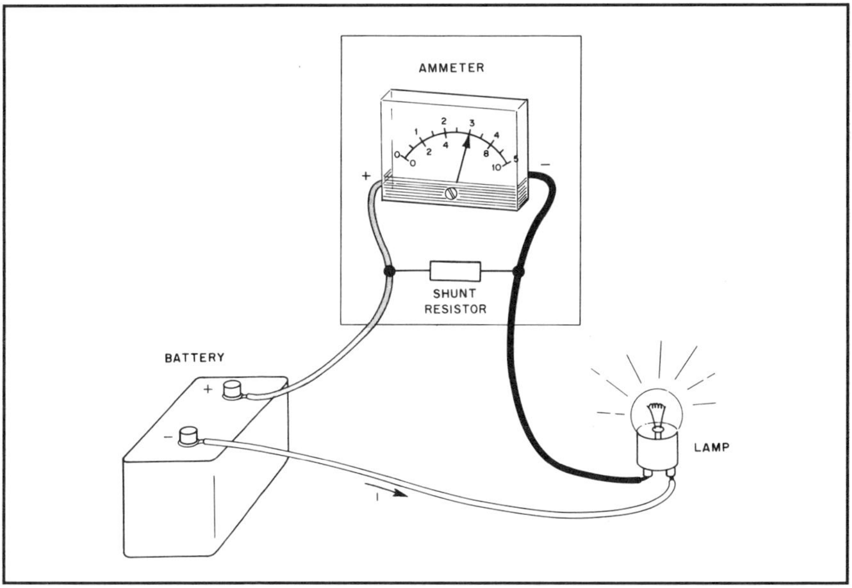

Figure 4-8—To measure current you must break the circuit at some point and connect the meter in series at the break. A shunt resistor expands the scale of the meter to measure higher currents than it could normally handle.

resistor values are calculated so the total circuit current can be read on the meter. Most ammeters have very low resistance, but low-impedance circuits require caution. There is a chance that the slight additional resistance of the series ammeter will disturb circuit operation.

Multimeters

A **multimeter** is a piece of test equipment that most amateurs should know how to use. The simplest kind of multimeter is the *volt-ohm-milliammeter (VOM)*. As its name implies, a VOM measures voltage, resistance and current. VOMs use one basic meter movement for all functions. The movement requires a fixed amount of current (often 1 mA) for a full-scale reading. As shown in the two previous sections, resistors are connected in series or parallel to provide the proper voltage or current meter reading. In a VOM, a switch selects various ranges for voltage, resistance and current measurements. This switch places high-value dropping resistors in series with the meter movement for voltage measurement. It connects low-value shunt resistors in parallel with the movement for current measurement. These parallel and series resistors extend the range of the basic meter movement.

If you are going to purchase a VOM, buy one with the highest ohms-per-volt rating that you can find. Stay away from meters rated under 20,000 ohms per volt if you can.

Measuring resistance with a meter involves placing the meter leads across the component or circuit you wish to measure. Make sure to select the proper resistance scale. The full-scale-reading multipliers vary from 1 to 1000 and higher. The scale is usually compressed on the higher end of the range. See Figure 4-9. For best accuracy, keep the reading in the lower-resistance half of the scale. On most meters this is the right-hand side. Thus, if you want to measure a resistance of about 5000 ohms, select the R × 1000 scale. Then the meter will indicate 5.

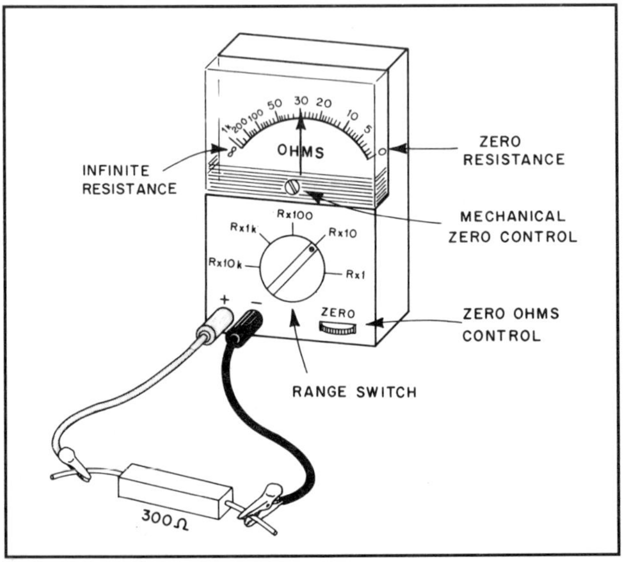

Figure 4-9—Resistance is measured across the component. For best accuracy the reading should be taken in the lower-resistance half of the scale.

A *vacuum-tube voltmeter (VTVM)* operates in the same manner as an ordinary VOM. There is one important difference. The meter in a VTVM is isolated from the circuit under test by a vacuum-tube dc amplifier. As a result, the only additional circuit loading is from the tube input impedance, which is very high. The standard VTVM input impedance is 11 MΩ. This is useful for measuring voltages in high-impedance circuits, such as vacuum-tube grid circuits and FET gate circuits.

Another type of meter is a *field-effect transistor volt-ohm-milliammeter (FET VOM)*. This instrument uses a field-effect transistor (FET) to isolate the indicating meter from the circuit to be measured. FET VOMs have an input impedance of several megohms. They are the solid-state equivalent of a VTVM.

[Turn to Chapter 10 now and study questions T4B01 through T4B06. Review this section if any of these questions give you difficulty.]

Wattmeters

A **wattmeter** is a device in the transmission line to measure the power (in watts) coming out of a transmitter. Wattmeters are designed to operate at a certain line impedance, normally 50 ohms. Make sure the feed-line impedance is the same as the design impedance of the wattmeter. If impedances are different, any measurements will be inaccurate. For the most accurate measurements, connect your wattmeter directly at the transceiver output (antenna) jack. See Figure 4-10.

All wattmeters must contain some type of detection circuitry to enable you to measure power. The accuracy and upper frequency range of the wattmeter is limited

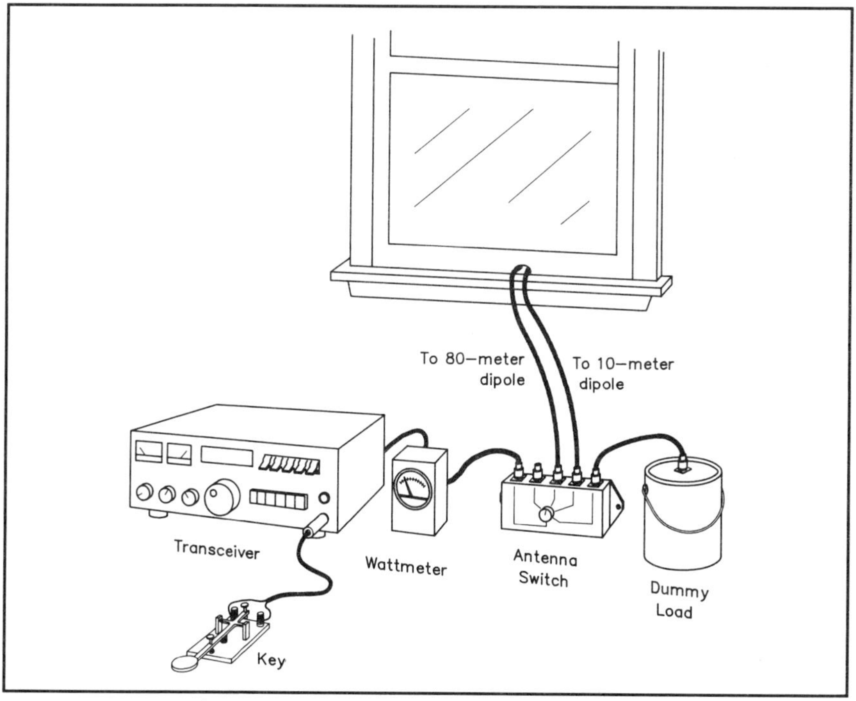

Figure 4-10—Connect a wattmeter at the transmitter output to measure output power.

by the detection device. Stray capacitance and coupling within the detection circuits can cause problems. A loss of sensitivity or accuracy occurs as the operating frequency is increased. A power meter designed for use at 3-30 MHz (HF) will be inaccurate at VHF or UHF.

One type of wattmeter is a **directional wattmeter**. There are two kinds of directional wattmeters. One has a meter that reads forward power and another meter that reads reflected power. The other has a single meter that can be switched to read either forward or reflected power.

You can use a directional wattmeter to measure the output power from your transmitter. Remember that the reflected power must be subtracted from the forward reading to give you the true forward power. The power reflected from the antenna will again be reflected by the transmitter. This power adds to the forward power reading on the meter. Let's say your transmitter power output is 100 watts and 10 watts of power is reflected from the antenna. Those 10 watts will be added into the forward reading on your meter when they are reflected from the transmitter. Your wattmeter reading would be incorrect. To find true forward power you must subtract the reflected measurement from the forward power measurement:

True forward power = Forward power reading – Reflected power reading (Equation 4-1)

For example, suppose you have a wattmeter connected in the line from your transmitter. It gives a forward-power reading of 85 watts and a reflected-power reading of 15 watts. What is the true power out of your transmitter?

True forward power = 85 W – 15 W = 70 watts.

[Now turn to Chapter 10 and study exam questions T4B07 through T4B11 as well as question T4C10. Review this section as needed.]

FREQUENCY ACCURACY

It's important to ensure that your transceiver frequency dial is accurate. If it isn't, you could transmit outside your allowed frequencies and be in violation of FCC rules. There are two good ways to ensure that your dial is accurate: use of a **marker generator** (**crystal calibrator**) and checking your dial by tuning in radio stations **WWV** or **WWVH**.

Marker Generators

The **marker generator** is a high-stability oscillator that produces an RF signal with a frequency that doesn't change. It is usually built into the receiver. This oscillator generates a series of signals that mark the exact edges of the amateur bands (and subbands, in some cases). It does this by oscillating at a low frequency that has harmonics falling on the desired frequencies. Most marker generators put out harmonics at 25, 50 or 100-kHz intervals. Since marker generators normally use crystal oscillators, they are often called **crystal calibrators**.

The marker generator is very useful. You can calibrate your receiver with it. You can then determine your precise transmitter frequency. First, turn on your marker generator (calibrator). This injects the calibrator signal into the receiver circuit. Then set your transceiver or receiver dial on the proper frequency marks. Calibration instructions are usually provided in your operator's manual. When the dial is calibrated, put your transmitter in the tune or spot position and read the frequency on the receiver dial.

The marker frequencies must be accurate. Now you know that the transmitter frequency is between the markers that show the ends of the band (or subband). In addition, the transmitter frequency must not be too close to the edge of the band (or subband). If it is, the sideband frequencies, especially in a voice transmission, will extend over the edge.

WWV and WWVH

Another easy way to calibrate your receiver frequency readout or dial is to tune in **WWV** (Fort Collins, Colorado) or **WWVH** (Kauai, Hawaii). These stations transmit reference signals on 1.5, 5, 10, 15 and 20 MHz. When you tune in WWV or WWVH, you'll hear a "tic" pulse every second and the precise time announced each minute.

Signal Generators

A **signal generator** produces a stable, low-level signal that can be set to a specific frequency. There are two different types of signal generators. Audio-frequency signal generators create signals in the audio range. Radio-frequency signal generators provide radio-frequency signals. You can use these signals to align or adjust tuned circuits for best performance. One common reason you might use a signal generator is to align a receiver. Set the generator to the desired signal frequency. Then you can adjust the associated circuits for best operation. You can usually use some type of meter to indicate the proper adjustment. For example, you could use the receiver's S meter to show when it is receiving a signal with maximum signal strength.

Sometimes you must cover a band of frequencies to chart the frequency response of a filter. For this application, you would use a swept frequency generator. A swept frequency generator automatically sweeps back and forth over a selected range of frequencies.

You can use a signal generator to adjust the filter circuits in your transmitter. When you do, a **dummy antenna** must be connected to the transmitter output to avoid radiating a signal. (Dummy antennas are discussed in the "Station Accessories" section.)

[Now turn to Chapter 10 and study questions T4C01 through T4C07. Review any material as necessary.]

ANTENNA SYSTEM MEASUREMENTS

A properly operating antenna system is essential for a top-quality amateur station. If you build your own antennas, you must be able to tune them for maximum operating efficiency. Even if you buy commercial antennas, you must make some tuning adjustments because of differences in mounting location. Height above ground and proximity to buildings and trees will have some effect on how your antenna operates. You should be able to monitor your antenna system periodically for signs of problems, and troubleshoot failures as necessary.

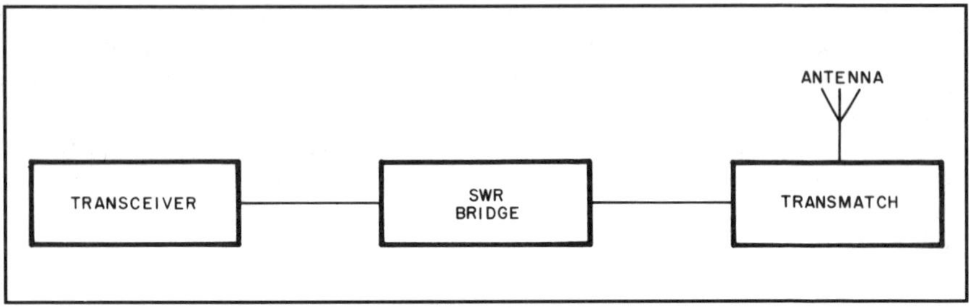

Figure 4-11—A reflectometer or SWR bridge should be connected between the transmitter and Transmatch in order to adjust the Transmatch properly.

Impedance Match Indicators

A **reflectometer** (**SWR meter**) is a useful metering device. A reflectometer is a device used to measure *standing-wave ratio*, or *SWR*. SWR is a measure of the relationship between the amount of power traveling to the antenna and the power reflected back to the transmitter. The reflection is caused by impedance mismatches in the antenna system. The energy going from the transmitter to the antenna is represented by the forward, or incident, voltage. The energy reflected by the antenna is represented by the reflected voltage.

A reflectometer is placed in the transmission line between the transmitter and antenna. In the most common amateur application, the reflectometer is connected between the transmitter and *Transmatch*, as shown in Figure 4-11. A Transmatch is used to cancel out the capacitive or inductive component in the antenna impedance. With a Transmatch you can match the impedance of your antenna system to the transmitter output, normally 50 ohms. The reflectometer is used to indicate minimum reflected power as the Transmatch is adjusted. This indicates when the antenna system (including feed line) is matched to the transmitter output impedance.

To obtain a valid measure of the impedance match between an antenna and the feed line, place the SWR meter at the antenna feed point. If you put the meter at the transmitter end, feed line losses make SWR readings look lower than they really are. Most SWR meters are designed for HF use (from 3 to 30 MHz). These may give a useful indication at VHF, if the meter properly calibrates to full scale in the set position. Don't place too much trust in the actual measured SWR values, however.

[Now turn to Chapter 10 and study questions T4C08, T4C09 and T4C11. Review the material in this section as necessary.]

STATION ACCESSORIES

Until now, we have talked about the bare necessities—what you need to set up a basic Amateur Radio station. Few hams are content with the bare necessities, however. Most add accessories that make operating more convenient and enjoyable. Let's look at some important station items: dummy antennas, and S meters.

Dummy Antenna

A **dummy antenna**, sometimes called a **dummy load**, is nothing more than a

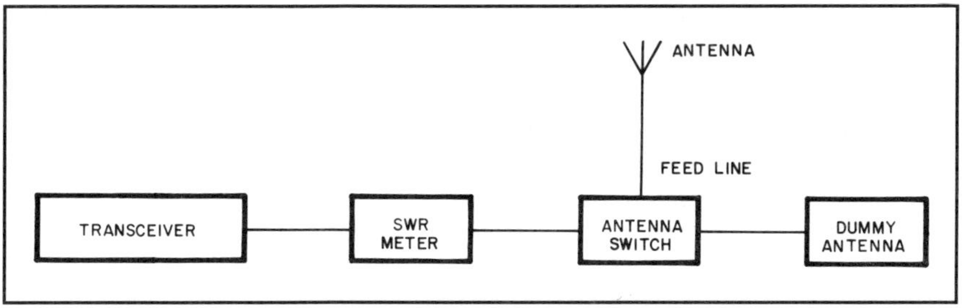

Figure 4-12—A dummy antenna is just a resistance that provides your transmitter with a proper load when you are tuning up and don't want to radiate a signal. Connect it to your antenna switch.

large resistor. It replaces your antenna when you want to operate your transmitter without radiating a signal. The dummy antenna safely converts the RF energy coming out of your transmitter into heat. The heat goes into the air or into the coolant, depending on the type of dummy antenna. It does all this while presenting your transmitter with a constant 50-ohm load. Dummy antennas are used to make test transmissions, or to make tuning adjustments after changing bands. Be sure the one you choose is rated for the type of operating (HF or VHF/UHF) and power level you'll be using.

Relatively inexpensive, a dummy antenna is one of the most useful accessories in your station. Every conscientious amateur should own one. You can build your own or buy one that's ready to use. The container, often a gallon paint can, acts as a shield to keep RF energy from being radiated. When filled with transformer oil, it allows the dummy antenna to dissipate more RF power. You can use an antenna switch to connect the dummy antenna to the transmitter. See Figure 4-12.

With a dummy antenna connected to the transmitter, you can make off-the-air tests—no signal goes out over the air. For transmitter tests, remember that a dummy antenna is a resistor. It must be capable of safely dissipating the entire (continuous) power output of the transmitter. A dummy antenna will often get warm during use, since the RF energy from the transmitter is turned into heat.

A dummy antenna should provide the transmitter with a perfect load. This usually means that it has a pure resistance (with no reactance) of approximately 50 ohms. The resistors used to construct dummy antennas must be noninductive. Composition resistors are usable, but wire-wound resistors are not. A single high-power resistor is best. Several lower-power resistors can be connected in parallel to obtain a 50-ohm load capable of dissipating high power.

The S Meter

S meters indicate the relative strength of a received signal. They are calibrated in S units from S1 to S9; above S9 they are calibrated in decibels (dB), as shown in Figure 4-13. Nearly every commercially built receiver and trans-

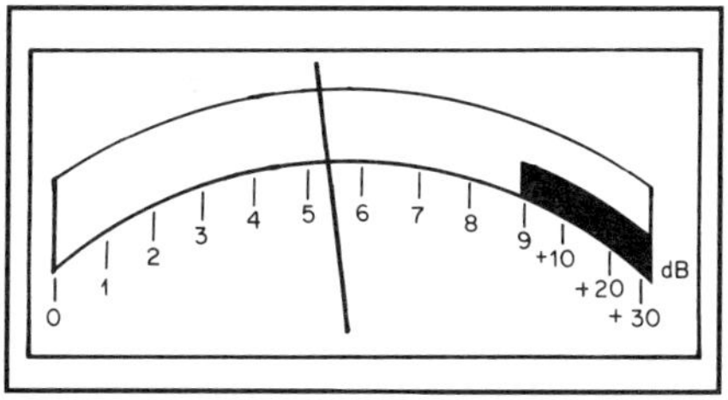

Figure 4-13—An S-meter is calibrated in S-units up to S-9. Above S-9 it is calibrated in decibels (dB) over S-9.

ceiver includes a built-in S meter. If you decide to build your own receiver, however, you may have to add an S meter to the basic design if you want one.

In the 1940s at least one manufacturer made an attempt to establish some significant numbers for S meters. S9 was to be equal to a signal level of 50 microvolts, with each S unit equal to 6 dB. This means that for an increase of one S unit, the received signal power would have to increase by a factor of four. Such a scale is useful for a theoretical discussion. Real S-meter circuits fall far short of this ideal for a variety of reasons.

Suppose you were working someone who was using a 25-watt transmitter, and your S meter was reading S7. If the other operator increased power to 100 watts (an increase of four times), your S meter would read S8. If the signal was S9, the operator would have to multiply his power 10-fold to make your S meter read 10 dB over S9. It would take another 10 times increase in signal strength to make the meter read 20 dB over S9.

S meters on modern transceivers may or may not respond in this manner. S-meter operation is based on the output of the *automatic gain-control (AGC) circuitry;* the S meter measures the AGC voltage. As a result, every S meter responds differently; no two S meters will give the same reading. The S meter is useful for giving *relative* signal-strength indications, however. You can see changes in signal levels on an S meter that you may not be able to detect just by listening to the audio output level.

[Now turn to Chapter 10 and study questions T4D01 through T4D08. Review the material in this section if you have any difficulty with these questions.]

SAFETY WITH RF POWER

Amateur Radio is basically a safe activity but accidents can always occur if we don't use common sense. Most of us know enough not to place an antenna where it can fall on a power line. We don't insert our hand into an energized linear amplifier, or climb a tower on a windy day. We also should not overexpose ourselves to RF energy. Large amounts of RF energy can be harmful to people because it heats body tissues. The effects depend on the wavelength, energy density of the incident RF field, and on other factors such as polarization.

Some of the most susceptible parts of the body are the tissues of the eyes. They don't have heat-sensitive receptors to warn us of the danger before the damage occurs. Symptoms of overexposure may not appear until after irreversible damage has been done. Though the problem should be taken seriously, with reasonable precaution, Amateur Radio operation can be safe.

Safe Exposure Levels

In recent years scientists have devoted a great deal of effort to determining safe RF-exposure limits. The problem is very complex. It's not surprising that some changes in the recommended levels have occurred as more information has become available. The American Radio Relay League believes that the 1982 **Radio Frequency Protection Guide of the American National Standards Institute (ANSI)** is a good protection standard. It took nearly five years to formulate and has undergone repeated critical review by the scientific community.

The **ANSI RF protection guide** recommends that special precautions be taken

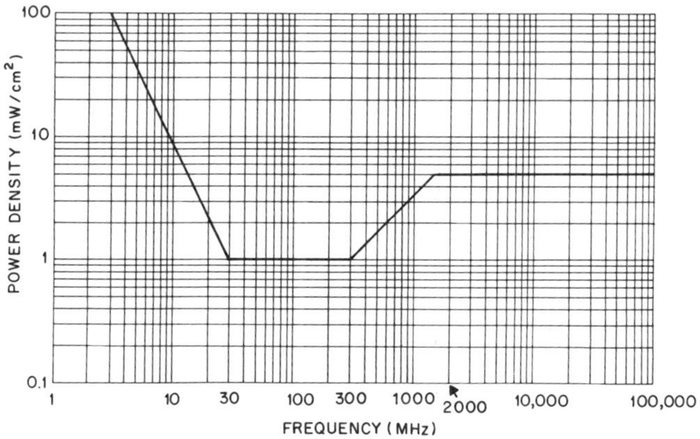

Figure 4-14—American National Standards Institute Radio Frequency Protection Guide for whole-body exposure of humans.

when transmitting with hand-held VHF and UHF radios. The antenna should be kept at least 1 to 2 inches away from the forehead, and power should be limited to no more than 7 watts.

The ANSI standards recognize the phenomenon of whole-body or geometric resonance and establishes a frequency-dependent maximum permissible RF exposure limit for humans. Exposure levels are expressed in terms of power density, measured in milliwatts per square centimeter. It is a measure of the radio-frequency power that strikes a person per square centimeter of body surface.

Whole-Body Resonance

Resonance occurs at frequencies for which the human body's length (height), if parallel to the antenna (vertical), is about 0.4 wavelength long. Because of the range of human heights, the resonant region spans a broad range of frequencies. This whole-body resonance establishes the frequency range for the most stringent (lowest) permissible exposure level. Figure 4-14 shows that the lowest maximum exposure level is 1 mW/cm^2 for frequencies between 30 and 300 MHz. The frequencies in this range cause the greatest potential risk for humans. RF energy at these frequencies is most easily absorbed by the human body. For this reason, it's important to reattach all shielding to a VHF amplifier after you've adjusted it. If it's turned on before all the shielding is in place, you could be exposed to dangerous amounts of RF energy.

On either side of those "corner" frequencies the rise is gradual. At 3 MHz the maximum permissible exposure level is 100 mW/cm^2; at 1500 MHz and above, 5 mW/cm^2. The valley region includes some active amateur bands (10, 6 and 2 meters) as well as all FM and TV broadcasting. The rationale for specifying a constant 5 mW/cm^2 above 1500 MHz takes into consideration that the extremely short wavelengths don't penetrate very deep into body tissue.

RF Safety Guidelines

Take the time to study and follow these general guidelines to minimize your exposure to RF fields. Most of these guidelines are just common sense and good amateur practice.

• Confine RF radiation to the antenna, where it belongs. Provide a good earth ground for your equipment. Particularly at VHF and UHF, poor-quality feed line and improperly installed connectors can be a source of unwanted RF radiation. Use only good-quality coaxial cable. Be sure the connectors are of good quality and are properly installed. Good-quality coaxial cable and connectors will also help reduce RF loses in your system.

• Don't operate RF power amplifiers or transmitters with the covers or shielding removed. This practice helps you avoid both electric shock hazards and RF safety hazards. A safety interlock prevents the gear from being turned on accidentally while the shielding is off. This is especially important for VHF and UHF equipment. When reassembling transmitting equipment, replace all the screws that hold the RF compartment shielding in place. Tighten all the screws securely before applying power to the equipment.

• In high-power operation in the HF and VHF region, keep the antenna away from people. Humans should not be allowed within 10 to 15 feet of vertical antennas. This is especially important with higher power, high-duty-cycle operation (such as FM or RTTY). Amateur antennas that are mounted on towers and masts, away from people, pose no exposure problem.

• Always install your antennas where people and animals cannot touch them.

• When using mobile equipment with 10-W RF power output or more, do not transmit if anyone is standing within 2 feet of the antenna.

• The best location for a VHF/UHF mobile antenna—from an RF safety standpoint and for the best radiated-signal pattern—is in the middle of the automobile roof. This position best protects the car's occupants.

• When using a hand-held transceiver with RF power output of several watts or more, maintain at least 1 to 2 inches separation between the antenna and your forehead to protect your eyes. It is recommended that hand-held radios have a power of no more than 7 watts.

• Never touch an antenna that has RF power applied. Be sure RF power is off and stays off before working on or adjusting an antenna. Also, make sure any nearby antennas are deactivated. Never have someone else transmit into the antenna and monitor the SWR while you are making adjustments. When matching an antenna, you should turn the transmitter off and make the adjustment. Then, back away to a safe distance before turning the transmitter on again to check your work.

• During transmissions, never point a high-gain UHF or microwave antenna (such as a parabolic dish) toward people or animals.

• Never look into the open end of a UHF or microwave waveguide feed line that is carrying RF power. Never point the open end of a UHF waveguide that is carrying RF power toward people or animals. Make sure that all waveguide connections are tightly secured.

The following additional safety guidelines were developed in response to scientific studies done in the mid to late 1980s. (From the October 1989 *QST* article, "Is Amateur Radio Hazardous to Our Health?" by Ivan A. Shulman, MD, WC2S.)

• Do not stand or sit close to your power supplies or linear amplifiers while operating, even when they are in standby mode.

• Stay at least 24 inches away from any power transformer, electrical fans or other source of high level 60-Hz magnetic fields while in operation.

• Do not tune up or operate a high-powered linear amplifier while the shields or covers are off.

• Run your transmission lines away from where you or other people sit in or near your shack.

• Properly terminated coaxial transmission feed lines should be used in preference to open-wire or end-fed antenna installations that come directly into the transmitter, as the RF radiated from a coaxial feed line is much lower.

• Use common sense about placing all antennas well away from yourself and others, especially for VHF, UHF and particularly microwave application. No one should be in the near field of an antenna.

• No person should be near any transmitting antenna while it is operating. This is especially true for mobile or ground-mounted vertical antennas. The use of indoor transmitting antennas that are close to people in a house or apartment should be reconsidered.

• Use the minimum power needed to make a QSO, especially if the antenna is less than 35 feet above the ground.

• Hand-held radios should be used on the lowest power setting needed to carry out communications.

• Hand-helds should be kept as far from the head as possible when operating. The use of a separate microphone or similar device is recommended.

• Transmissions using a hand-held radio should be kept as short as possible.

• Power density measurements should be made before running more than 25 watts in a VHF mobile installation, particularly if the antenna is rear-deck mounted and passengers may ride in the back seat. The safest mobile antenna location is in the center of the metal roof.

• The development of an accurate inexpensive power-density meter would be of major benefit to the Amateur Radio community so that RF power-density measurements could be taken in all radio installations. Because of the current high cost of such devices, groups of amateurs or clubs may wish to purchase one and share in its use.

• Soldering should only be done in a well ventilated area. A small fan should be used to blow away toxic fumes.

• When using toxic chemicals, such as when etching PC boards or repairing fiberglass, wear gloves and goggles, use proper tools, and avoid contact with any of the chemicals. If accidentally contaminated, wash off the compounds immediately with lots of water. Again, the importance of always working in a well ventilated area with personal protective covering cannot be overemphasized.

• Hazardous chemicals, such as those in the PCB class, are used in some capacitors and dummy loads. Use extreme care in handling these materials, and consult with the appropriate local authorities to determine the proper means of disposing of these chemicals in an environmentally responsible way.

[This completes your study of Chapter 4. Now turn to Chapter 10 and study questions T4D09 through T4D16. Review the material in this section if you have any difficulty with these questions.]

KEY WORDS

Capacitance—The ability of a capacitor to store energy in an electric field.

Capacitor—An electrical component usually formed by separating two conductive plates with an insulating material. A capacitor stores energy in an electric field.

Farad—The basic unit of capacitance.

Henry—The basic unit of inductance.

Induced EMF—A voltage produced by a change in magnetic lines of force around a conductor. When a magnetic field is formed by current in the conductor, the induced voltage always opposes change in that current.

Inductance—The ability of a coil to store energy in a magnetic field.

Inductor—An electrical component usually composed of a coil of wire wound on a central core. An inductor stores energy in a magnetic field.

Ohm—The basic unit of electrical resistance, used to describe the amount of opposition to current.

Ohm's Law—A basic law of electronics. Ohm's Law gives a relationship between voltage (E), current (I) and resistance (R). The voltage applied to a circuit is equal to the current through the circuit times the resistance of the circuit ($E = IR$).

Parallel circuit—An electrical circuit in which the electrons follow more than one path in going from the negative supply terminal to the positive terminal.

Resistance—The ability to oppose an electric current.

Resistor—Any material that opposes a current in an electrical circuit. An electronic component especially designed to oppose current.

Series circuit—An electrical circuit in which all the electrons must flow through every part of the circuit. There is only one path for the electrons to follow.

5 Electrical Principles

To pass your FCC Element 2 written exam and receive your Novice license, you had to learn some very basic radio theory. For your Element 3A (Technician) test, you will need to understand a few more electrical principles. This chapter will introduce you to the basic concepts of resistance, capacitance and inductance.

Ohm's Law is an important tool for analyzing circuits. We'll look at some applications of this basic electronics principle. We'll look at methods for finding the total value of parallel and series combinations of resistors, capacitors and inductors.

Be sure to turn to Chapter 10 and study the appropriate exam questions when the text tells you to do so. This will show you if you are progressing smoothly, or if you need a little extra studying. We cannot tell you everything about a particular topic in one chapter. You may want to use some other reference books for additional information. The ARRL's *Understanding Basic Electronics* is a good starting point. *The ARRL Handbook for Radio Amateurs* contains more detailed information.

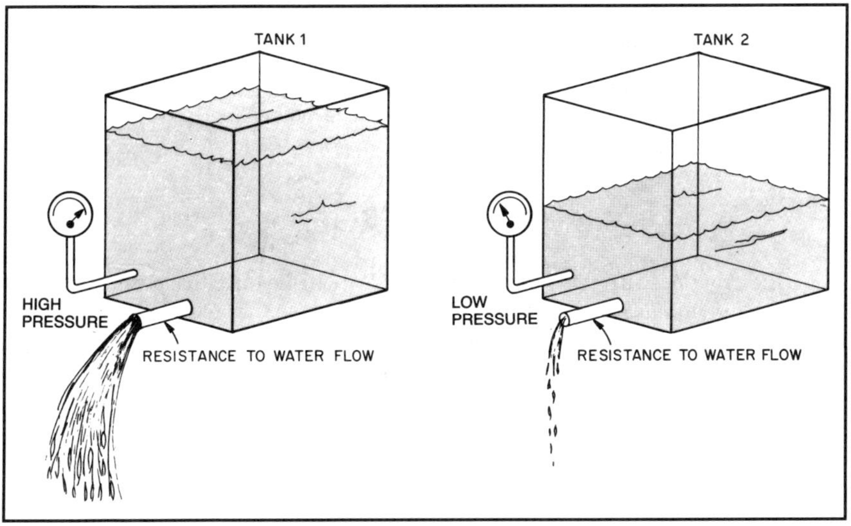

Figure 5-1—Since the water-tank outlet pipes are identical in size, they present an identical amount of resistance to water flow. The pressure at the outlets is different because of the difference in water level. This is similar to two separate electrical circuits with identical amounts of resistance and different supply voltages. The higher pressure of tank 1 causes more water to flow over a given time, just as a higher voltage causes larger electrical currents.

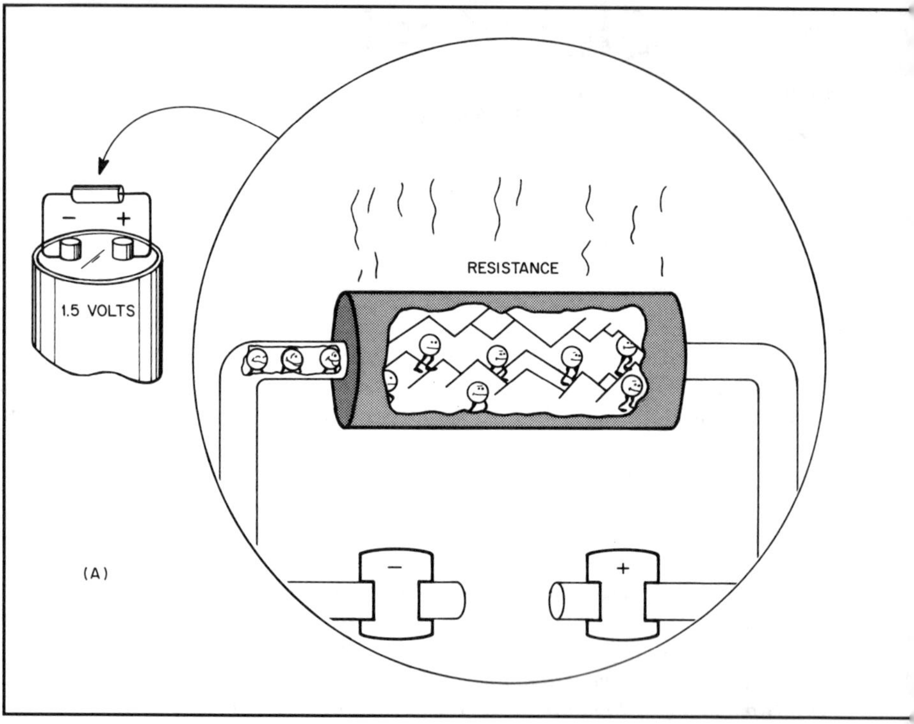

Figure 5-2—At A, 1.5 volts supplies electron in the wire and resistive material with energy to try to "push" their way through the circuit from one battery terminal to the other. The number of electrons that can move in a given amount of time is limited by resistance. The electrons find it easy going in the low-resistance wire, but have to expend more energy in passing through the higher resistance. This energy is released as heat. The 12-volt battery at B has the electrical force needed to move many more electrons. Much more heat is given off by the resistive material.

E = VOLTAGE

Electrons need a pus to get them moving. We call the force that pushes electrons through a circuit *electromotive force (EMF)*. EMF is similar to water pressure in a pipe. See Figure 5-1. More pressure causes more water to flow through the pipe. Greater EMF causes more electrons to flow through the circuit.

EMF is measured in volts, so it is usually called *voltage*. A voltage-measuring instrument is a voltmeter. More voltage applied to a circuit makes more electrons flow through the circuit. See Figure 5-2. We use the letter E to represent EMF or voltage in equations and on diagrams.

I = CURRENT

The word *current* comes to us from a Latin word meaning "to run" and it implies movement. When someone speaks of the current in a river, we think of a flow of water.

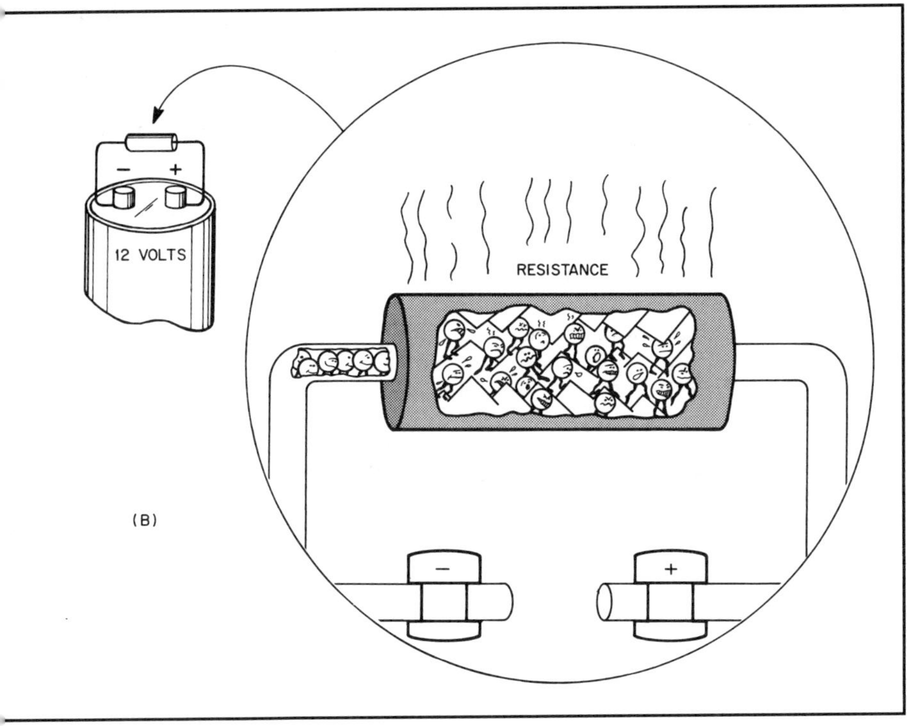

A current of air can be a light breeze or a hurricane. A current of electricity is a flow of electrons.

Current is indicated in equations and diagrams by the letter I (from the French word *intensité*). We measure current in amperes, and the measuring device is an ammeter. An ampere is often an inconveniently large unit. We can measure current in milliamperes (one thousandth of an ampere) or microamperes (one millionth of an ampere). The abbreviation for amperes (sometimes referred to informally as *amps*) is A; for milliamperes we use mA and for microamperes μA.

R = RESISTANCE

Resistance is something that opposes motion. All materials have electrical resistance. There is always some opposition to the flow of electrons through the material. The resistance of different materials varies widely. Some materials have very high resistance; we call these materials good insulators, or poor conductors. Other materials have very low resistance; we call these materials poor insulators, or good conductors. The basic unit of resistance is the **ohm**, and a resistance-measuring instrument is an ohmmeter. The symbol for an ohm is Ω, the Greek letter capital omega. We use the letter R to represent resistance in equations and on diagrams.

We can make devices with a range of opposition to electron flow in a circuit. They are between the extremes of a very good insulator and a very good conductor. We call such devices **resistors**. We will use the analogy of water flowing in a pipe to help us understand electron flow. These materials act like a pipe with a sponge in it. Some water will still flow in the pipe, but the flow will be reduced. The primary function of a resistor is to limit electron flow (current). Greater resistance reduces the current even more. If we want to control the current reduction over a certain range, we can use a variable resistor. A variable resistor can be adjusted over a range of resistance. We can then control the flow of electrons from a maximum to a minimum over this range.

This current reduction has a price, however. The energy lost by the electrons as they flow through a resistor must be dissipated in some way. The energy converts into heat, and the resistor gets warm. if the current is higher than the resistor can handle, the resistor will get very hot and may even burn.

[Before you go on to the next section, turn to Chapter 10 and study question T5A01. Review this section if you have any difficulty with this question.]

Resistors in Series and Parallel

Sometimes it's necessary to calculate the total resistance of resistors connected in a **series circuit**. Resistors in series are connected end to end like a string of sausages. At other times you must calculate total resistance in a parallel circuit. In a parallel circuit, resistors are side by side like a picket fence. There will be times when you need a certain amount of resistance somewhere in a circuit. There may be no standard resistor value that will give the necessary resistance. Sometimes you may not have a certain value on hand. By combining resistors in parallel or series you can obtain the desired value.

When you connect resistors in series, as in Figure 5-3, the total resistance is simply the sum of all the resistances. Let's use the analogy of a sponge in a water pipe again. Connecting resistors in series is like putting several sponges in the same pipe. The total resistance to the water would be the sum of all the individual resistances. Resistors in series add.

$$R_{TOTAL} = R_1 + R_2 + R_3 + \ldots + R_n \qquad \text{(Equation 5-1)}$$

where n is the total number of resistors.

The total resistance of a string of resistors in series will always be greater than any individual resistance in the string. All the circuit current flows through each resistor in a series circuit. A series circuit with resistor values of 2 ohms, 3 ohms and 5 ohms would have a total resistance of 10 ohms. If a circuit has two equal-value resistors connected in series, the total resistance will be twice the value of either resistor alone.

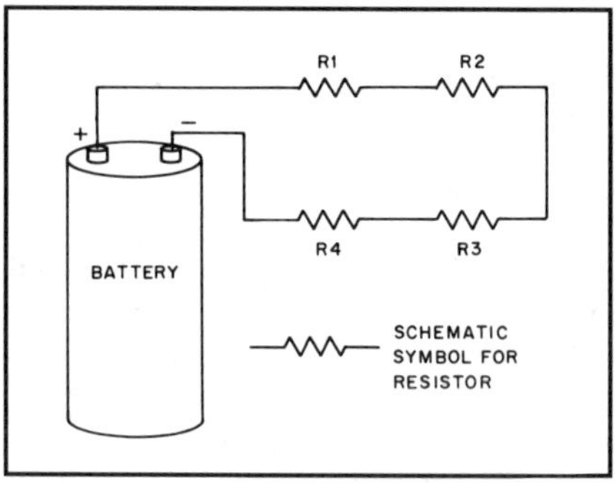

Figure 5-3—The total resistance of a string of series-connected resistors is the sum of all the individual resistances.

In a **parallel circuit**, things are a bit different. When we connect two or more resistors in parallel, more than one path for current exists in the circuit. See Figure 5-4. This is like connecting another pipe into our water-pipe circuit.

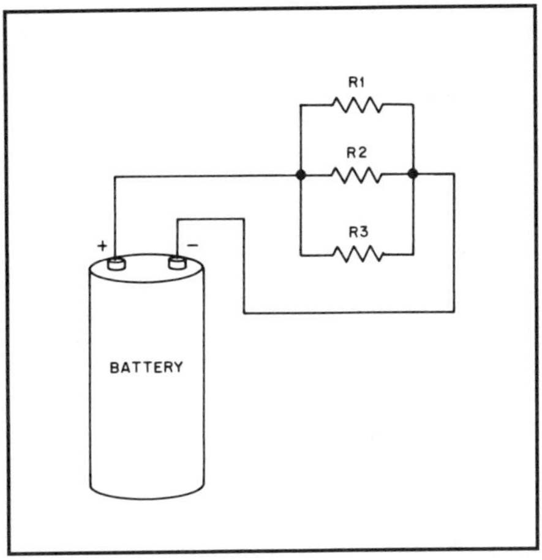

Figure 5-4—In this circuit, the electrical current splits into three separate paths through the resistors. Each resistor is exposed to the full battery voltage. Current through each individual resistor is independent of the other resistors. For example, if R_2 and R_3 are changed to a value different than the original value, the current in R_1 remains unchanged. Even if R_2 and R_3 are removed, the current in R_1 remains unchanged.

When there is more than one path, more water can flow during a given time. With more than one resistor, more electrons can flow. This means there is less resistance and a greater current.

The formula for calculating the total resistance of resistors connected in parallel is:

$$R_{TOTAL} = \frac{1}{\frac{1}{R_1} + \frac{1}{R_2} + \frac{1}{R_3} + \cdots + \frac{1}{R_n}} \qquad \text{(Equation 5-2)}$$

where n is the total number of resistors.
For example, if we connect three 100-ohm resistors in parallel, their total resistance is:

$$R_{TOTAL} = \frac{1}{\frac{1}{100} + \frac{1}{100} + \frac{1}{100}}$$

$$R_{TOTAL} = \frac{1}{\frac{3}{100}} = \frac{1}{0.03} = 33.33 \text{ ohms}$$

If we connect a 25-ohm resistor, a 100-ohm resistor and a 10-ohm resistor in parallel, our equation becomes:

$$R_{TOTAL} = \frac{1}{\frac{1}{25} + \frac{1}{100} + \frac{1}{10}}$$

$$R_{TOTAL} = \frac{1}{0.04 + 0.01 + 0.1} = \frac{1}{0.15} = 6.67 \text{ ohms}$$

To calculate the total resistance of two resistors in parallel, Equation 5-2 reduces to the "product over sum" formula. Divide the product of the two resistances by their sum:

$$R_{TOTAL} = \frac{R_1 \times R_2}{R_1 + R_2} \qquad \text{(Equation 5-3)}$$

If we connect two 50-ohm resistors in parallel, the total resistance would be:

$$R_{TOTAL} = \frac{50 \times 50}{50 + 50} = \frac{2500}{100} = 25 \text{ ohms}$$

The total resistance of two equal-value resistors connected in parallel is half the value of one of the resistors.

Now let's look at any parallel combination of resistors. The total resistance is always less than the smallest value of the parallel combination. You can use this fact to make a quick check of your calculations. The result you calculate should be smaller than the smallest value in the parallel combination. If it isn't, you've made a mistake somewhere!

OHM'S LAW

The amount of water flowing through a pipe increases as we increase the pressure and decreases as we increase the resistance. If we replace "pressure" with "voltage," we can write a mathematical relationship for an electric circuit:

$$\text{current} = \frac{\text{voltage}}{\text{resistance}} \qquad \text{(Equation 5-4)}$$

This equation tells us the current through a circuit equals the voltage applied to the circuit divided by the circuit resistance.

If the voltage stays constant but more current flows in the circuit, we know there must be less resistance. The relationship between current and voltage is a measure of the resistance:

$$\text{resistance} = \frac{\text{voltage}}{\text{current}} \qquad \text{(Equation 5-5)}$$

We can state this equation in words as: The circuit resistance is equal to the voltage applied to the circuit divided by the current through the circuit.

Finally, we can determine the voltage if we know how much current is flowing and the resistance in the circuit:

$$\text{voltage} = \text{current} \times \text{resistance} \qquad \text{(Equation 5-6)}$$

The voltage applied to a circuit is equal to the current through the circuit times the circuit resistance.

Scientists are always looking for shorthand ways of writing these relationships. They use symbols to replace the words: E represents voltage (remember EMF?), current is I (from the French word *intensité*) and resistance is R. We can now express **Ohm's Law** in a couple of letters:

$$E = IR \text{ (volts = amperes} \times \text{ohms)} \qquad \text{(Equation 5-7)}$$

This is the most common way to express Ohm's Law, but we can also write it as:

$$I = \frac{E}{R} \text{ (amperes = volts divided by ohms)} \qquad \text{(Equation 5-8)}$$

and

$$R = \frac{E}{I} \text{ (ohms = volts divided by amperes)} \qquad \text{(Equation 5-9)}$$

E is EMF in volts, I is the current in amperes and R is the resistance in ohms. If you know two of the numbers, you can calculate the third. Figure 5-5 shows a diagram to help you solve Ohm's Law problems. Simply cover the symbol of the quantity that you do not know. If the remaining two are side-by-side, you must multiply them. If one symbol is above the other, then you must divide the quantity on top by the one on the bottom.

If you know current and resistance in a circuit, Ohm's Law will give you the voltage (Equation 5-7). For example, what is the voltage applied to the circuit if 5 amperes of current flows through 20 ohms of resistance? From Equation 5-7 or Figure 5-5, we see that we must multiply 5 amperes times 20 ohms to get the answer, 100 volts. The EMF in this circuit is 100 volts.

$$E = I\,R \qquad \text{(Equation 5-7)}$$

$$E = 5 \text{ amperes} \times 20 \text{ ohms}$$

$$E = 100 \text{ volts}$$

Figure 5-5—The "Ohm's Law Circle." This simple diagram will help you remember the Ohm's Law relationships. To find any quantity if you know the other two, simply cover the unknown quantity with your hand or a piece of paper. The positions of the remaining two symbols show if you have to multiply (if they are side by side) or divide (if they appear one over the other as a fraction). For example, if you cover the I, you are left with E / R.

Suppose you know voltage and resistance — 100 volts in the circuit to push electrons against 20 ohms of resistance? Equation 5-8 gives the correct equation, or you can use the diagram in Figure 5-5. You must divide 100 volts by 20 ohms to find that 5 amperes of current is flowing.

$$I = \frac{E}{R} \qquad \text{(Equation 5-8)}$$

$$I = \frac{100 \text{ volts}}{20 \text{ ohms}}$$

$$I = 5 \text{ amperes}$$

If you know voltage and current in a circuit, you can calculate resistance. Using our example, 100 volts is pushing 5 amperes through the circuit. This time Equation 5-9 is the one to use, and you can also find this from Figure 5-5. 100 divided by 5 equals 20, so the resistance is 20 ohms.

$$R = \frac{E}{R}$$

$$R = \frac{100 \text{ volts}}{5 \text{ amperes}}$$

$$R = 20 \text{ ohms}$$

If you know E and I, you can find R. If you know I and R, you can calculate E. If E and R are known, you can find I.

Put another way, if you know volts and amperes, you can calculate ohms. If

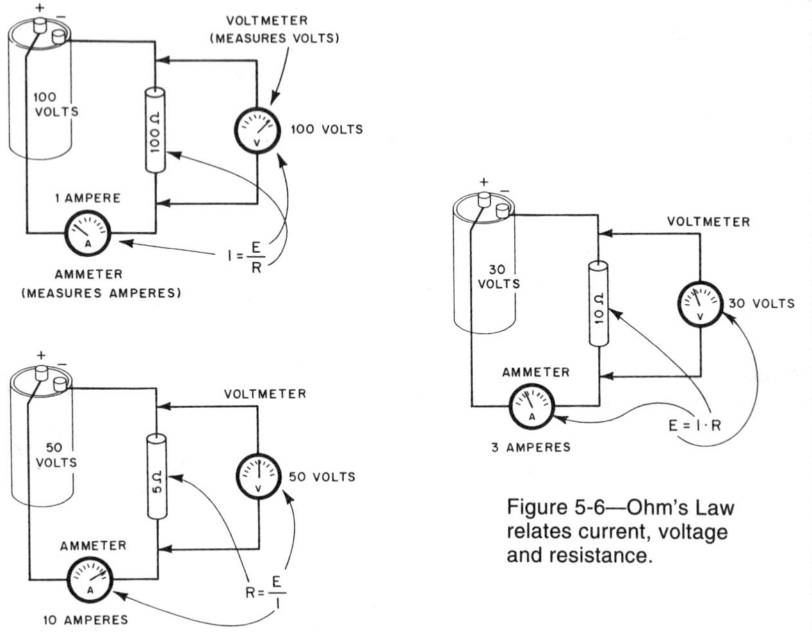

Figure 5-6—Ohm's Law relates current, voltage and resistance.

amperes and ohms are known, volts can be found. Or if volts and ohms are known, amperes can be calculated. Figure 5-6 illustrates some simple circuits and how Ohm's Law can be used to find an unknown quantity in the circuit. Make up a few problems of your own and test how well you understand this basic law of electricity. You'll soon find that this predictable relationship, symbolized by the repeatable equation of Ohm's Law, makes calculating values of components in electrical circuits easy!

[Before you read further, turn to Chapter 10 and study questions T5B01 through T5B11. Review the material in this section if you have difficulty with any of these questions.]

INDUCTANCE

The motion of electrons produces magnetism. Every electric current creates a *magnetic field* around the wire in which it flows. (A magnetic field represents the invisible magnetic force, such as the attraction and repulsion between magnets.) Like an invisible tube, the magnetic field is positioned in concentric circles around the conductor. See Figure 5-7A. The field is established when the current flows, and collapses back into the conductor when the current stops. The field increases in strength when the current increases and decreases in strength as the current decreases. The force produced around a straight piece of wire by this magnetic field is usually negligible. When the same wire is formed into a coil, the force is much greater. In coils, the magnetic field around each turn also affects the other turns. Together, the com-

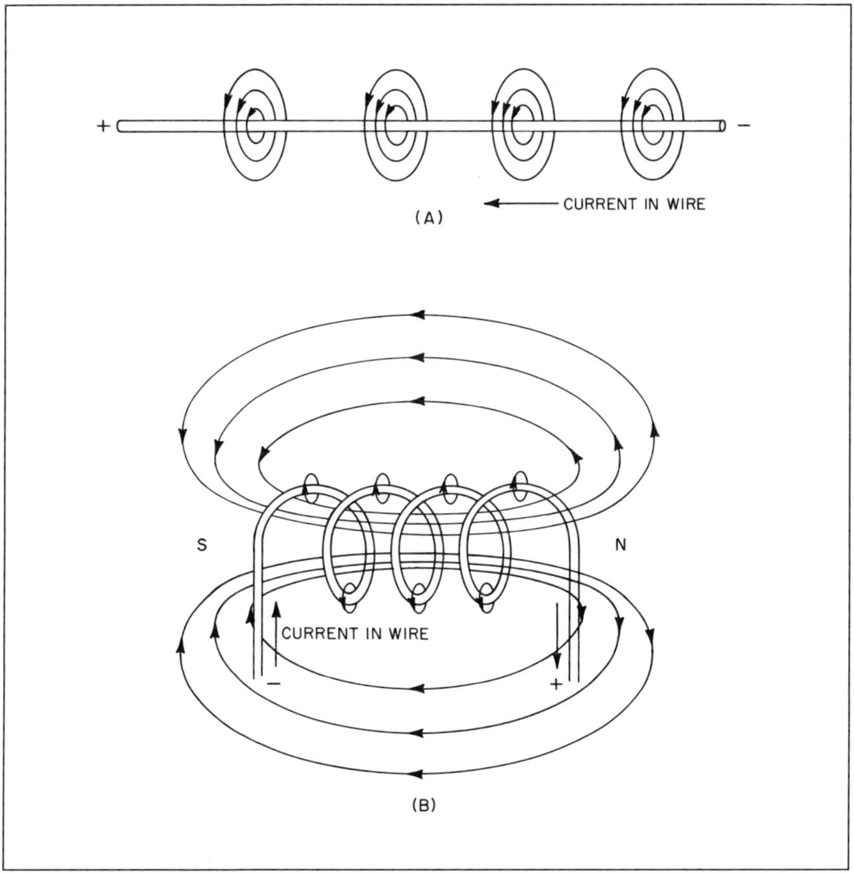

Figure 5-7—A magnetic field surrounds a wire with a current in it. If the wire is formed into a coil, the magnetic field becomes much stronger as the lines of force reinforce each other.

bined forces produce one large magnetic field, as shown in Figure 5-7B. Much of the energy in the magnetic field concentrates in the material in the center of the coil (the *core*). Most practical **inductors** consist of a length of wire wound on a core.

An inductor stores energy in a magnetic field. This property of a coil to store energy in a magnetic field is called **inductance**. Magnetic fields can also set electrons in motion. When a magnetic field increases in strength, the voltage on a conductor within that field increases. When the field strength decreases, so does the voltage.

Let's apply a dc voltage to an inductor. As there is no current to start with, the current begins to increase when the voltage is applied. This will establish an electrical current in the inductor. The voltage induced by the magnetic field of the inductor opposes the applied voltage. Therefore, the inductor will oppose the increase in current. This is a basic property of inductors. Any changes in current through the inductor, whether increasing or decreasing, are opposed. A voltage is induced in the coil that opposes the applied voltage, and tries to prevent the current from changing. This is called **induced EMF** (voltage) or back EMF. Gradually a current will be produced by the applied voltage. (The term "gradually" is relative. In radio circuits, the time needed to produce the current in the circuit is often measured in microseconds.)

The "final" current that flows through the inductor is limited only by any resistance that might be in the circuit. There is very little resistance in the wire of most coils. The current will be quite large if there is no other resistance.

In the process of getting this current to flow, energy is stored. This energy is in the form of a magnetic field around the coil. When the applied current is shut off, the magnetic field collapses. The collapsing field returns energy to the circuit as a momentary current that continues to flow in the same direction as the original current. This current can be quite large for large values of inductance. The induced voltage can rise to many times the applied voltage. This can cause a spark to jump across switch or relay contacts when the circuit is broken to turn off the current. This effect is called "inductive kickback."

When an ac voltage is applied to an inductor, the current through the inductor will reverse direction every half cycle. This means the current will be constantly changing. The inductor will oppose this change. Energy is stored in the magnetic field while the current is increasing during the first half cycle. This energy will be returned to the circuit as the current starts to decrease. A new magnetic field will be produced during the second half cycle. The north and south poles of the field will be the reverse of the first half cycle. The energy stored in that field will be returned to the circuit as the current again starts to decrease. Then a new magnetic field will be produced on the next half cycle. This process keeps repeating, as long as the ac voltage is applied to the inductor.

Factors That Determine Inductance

The inductance of a coil determines several circuit conditions. One is the amount of opposition to changes in current. Another is the amount of energy stored in the magnetic field. The back EMF induced in the coil also depends on the inductance. In turn, the inductance of a coil, usually represented by the symbol L, depends on four things:

1) the type of material used for the core (permeability of the material), and its size and location in the coil;
2) the number of turns used to wind the coil;
3) the length of the coil (spacing of turns);
4) the diameter of the coil (cross-sectional area).

Changing any of these factors changes the inductance. See Figure 5-8.

The basic unit of inductance is the **henry**, named for the American physicist Joseph Henry. The henry is often too large for practical use in measurements. We use the millihenry (10^{-3}) abbreviated mH, or microhenry (10^{-6}) abbreviated µH.

Inductors In Series and Parallel

In circuits, inductors combine like resistors. The total inductance of several inductors connected in series is the sum of all the inductances:

$$L_{TOTAL} = L_1 + L_2 + L_3 + \ldots + L_n \quad \text{(Equation 5-10)}$$

where n is the total number of inductors.

For two equal-value inductors connected in series, the total inductance will be twice the value of one of the inductors.

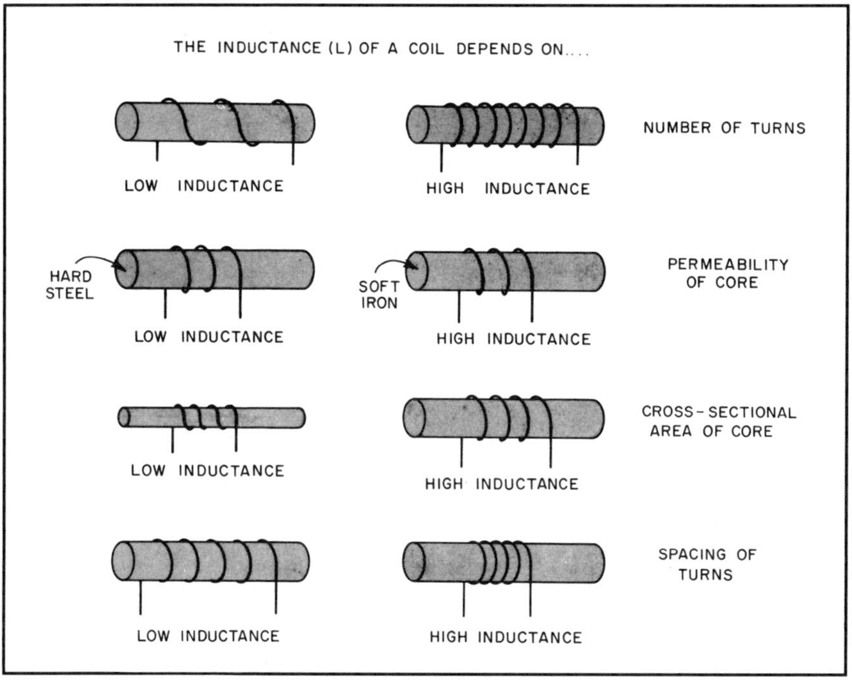

Figure 5-8—The value of an inductor depends on the material used in the core, the number of windings, and the length and diameter of the coil.

For parallel-connected inductors:

$$L_{TOTAL} = \frac{1}{\frac{1}{L_1} + \frac{1}{L_2} + \frac{1}{L_3} + \ldots + \frac{1}{L_n}} \qquad \text{(Equation 5-11)}$$

You should recognize this equation from our study of combining resistors in parallel, and realize that for two parallel-connected inductors, Equation 5-11 reduces to:

$$L_{TOTAL} = \frac{L_1 \times L_2}{L_1 + L_2} \qquad \text{(Equation 5-12)}$$

For two equal-value inductors connected in parallel, the total inductance will be half the value of one of the components.

[Turn to Chapter 10 and study questions T5A02, T5A03, T5A04, T5A08 and T5A09. Review this section as needed.]

CAPACITANCE

A simple **capacitor** is formed by separating two conductive plates with an insulating material. Connect one plate to the positive terminal of a voltage source. Connect the other plate to the negative terminal. We can build up a surplus of electrons on one plate, as shown in Figure 5-9. At some point, the voltage across the capacitor

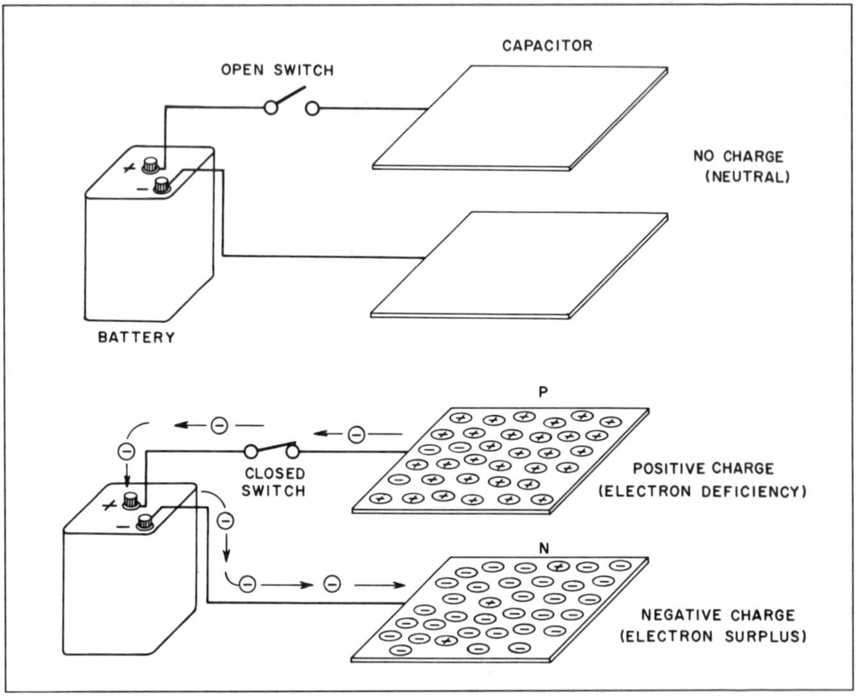

Figure 5-9—When a voltage is applied to a capacitor, an electron surplus (negative charge) builds up on one plate, while an electron deficiency forms on the other plate to produce a positive charge.

will equal the applied voltage, and the capacitor is said to be charged. This stored electric charge produces an *electric field*, which is an invisible electric force of attraction or repulsion acting between charged objects. (In this case the electric field is between the capacitor plates.) The capacitor stores energy as an electric field between the capacitor plates.

If we connect a load to a charged capacitor, it will discharge through the load, releasing stored energy. The basic property of a capacitor (called **capacitance**) is this ability to store a charge in an electric field.

The basic unit of capacitance is the **farad**, named for Michael Faraday. Like the henry, the farad is usually too large a unit for practical measurements. For convenience, we use microfarads (10^{-6}), abbreviated µF, or picofarads (10^{-12}), abbreviated pF.

Factors That Determine Capacitance

The capacitance value of a capacitor is determined by three factors. Increased plate-surface area will increase capacitance. Increased spacing between plates reduces capacitance. The type of insulating material (dielectric) used between the plates will affect capacitance. See Figure 5-10.

Reducing the spacing between plates or using a better insulator will increase the capacitance. For a given plate size, using multiple plates in the construction of a

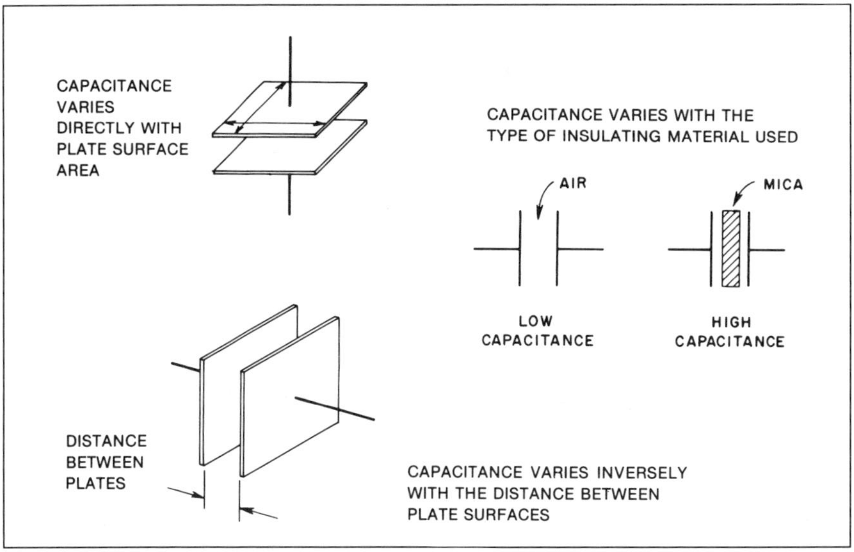

Figure 5-10—The capacitance of a capacitor depends on the area of the plates, the distance between the plates and the type of dielectric material used.

capacitor increases capacitance (by giving the effect of increased plate size). The effects of these factors are covered in more detail in Chapter 6.

Capacitors in Series and Parallel

We can increase the value of capacitance by increasing the total plate area. We can effectively increase the total plate area by connecting two capacitors in parallel. The two parallel-connected capacitors act like one larger capacitor, as shown in Figure 5-11A and B. The total capacitance of several parallel-connected capacitors is the sum of all the values.

$$C_{TOTAL} = C_1 + C_2 + C_3 + \ldots + C_n \qquad \text{(Equation 5-13)}$$

where n is the total number of capacitors.

If you connect two equal-value capacitors in parallel, the total capacitance of this combination will be twice the value of either capacitor alone.

Connecting capacitors in series has the effect of increasing the distance between the plates, thereby reducing the total capacitance, as shown in Figure 5-11C and D. For capacitors in series, we use the familiar reciprocal formula:

$$C_{TOTAL} = \frac{1}{\frac{1}{C_1} + \frac{1}{C_2} + \frac{1}{C_3} + \ldots + \frac{1}{C_n}} \qquad \text{(Equation 5-14)}$$

When there are only two capacitors in series, Equation 5-14 reduces to the product over sum formula:

$$C_{TOTAL} = \frac{C_1 \times C_2}{C_1 + C_2} \qquad \text{(Equation 5-15)}$$

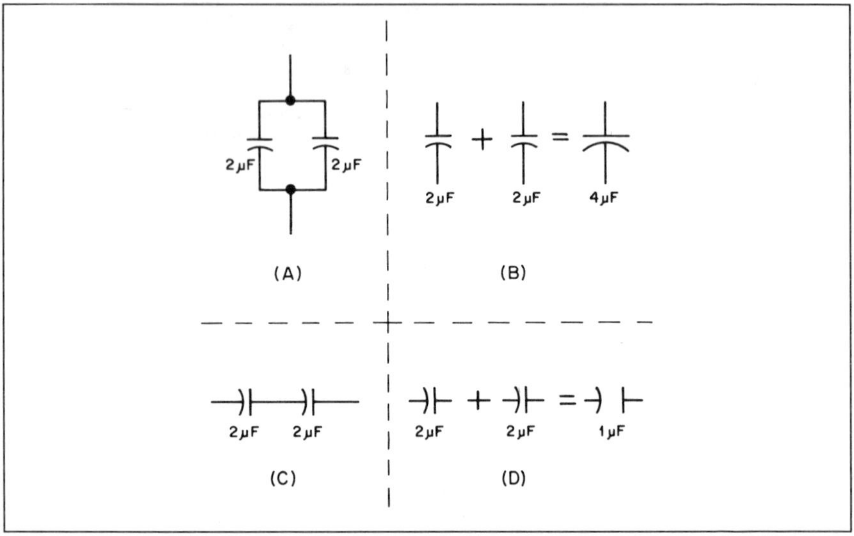

Figure 5-11—Parallel-connected capacitors are shown at A. This connection has the effect of increasing the total plate area, as shown at B. This increases the capacitance. Series connection, shown at C, has the effect of increasing the spacing between the plates, as shown at D. This decreases the capacitance.

Using this formula for two equal-value resistors or two equal-value inductors in parallel, we discovered the total value is half the value of either one. Two equal-value capacitors connected in series total one-half the value of either single capacitor.

[This completes your study of Chapter 5. Now turn to Chapter 10 and study exam questions T5A05, T5A06, T5A07, T5A10 and T5A11. Before proceeding to the next chapter, review the material in this section if you have difficulty with any of these questions.]

KEY WORDS

Capacitor—An electronic component composed of two or more conductive plates separated by an insulating material.

Color code—A system in which numerical values are assigned to various colors. Colored stripes are painted on the body of resistors and sometimes other components to show their value.

Core—The material used in the center of an inductor coil.

Electrolytic capacitor—A polarized capacitor that relies on a chemical reaction to form a thin insulating layer between the plates.

Electric field—An invisible force of nature. An electric field exists in a region of space if an electrically charged object placed in the region is subjected to an electrical force.

Film resistor—A resistor made by depositing a thin layer of resistive material on a ceramic form.

Inductor—An electrical component usually composed of a coil of wire wound on a central core.

Potentiometer—A resistor whose resistance can be varied continuously over a range of values.

Reactance—The property of an inductor or capacitor (measured in ohms) that impedes current in an ac circuit without converting power to heat.

Resistor—Any material that opposes a current in an electrical circuit. An electronic component specifically designed to oppose current.

Toroidal core—A donut-shaped form used to hold the turns of an inductor. Toroidal cores are usually made from ferrite or powdered-iron materials.

Variable capacitor—A capacitor that can have its value changed within a certain range.

Variable resistor—A resistor whose value can be adjusted over a certain range.

Wire-wound resistor—A resistor made by winding a length of wire on an insulating form.

6 Circuit Components

You can understand the operation of most complex electronics circuits. First, you must know some basic information about the parts that make up those circuits. This chapter presents the information about circuit components that you will need to know to pass your Technician class license exam. You will find descriptions of resistors, capacitors and inductors. You can combine these components with other devices to build practical electronic circuits. We describe some of these circuits in Chapter 7.

Study the characteristics of the components described in this chapter. Also study and be able to identify the schematic-diagram symbols for the components. (A schematic diagram shows component connections without regard to physical position.) Then turn to Chapter 10, and study the questions as you are directed to do throughout the chapter. That will help you check your understanding of the material and show you where you need to do some extra studying. You should thoroughly understand how these components work. When you do, you can connect them to make a circuit perform a specific task.

RESISTORS

When electrons flow from one point in a circuit to another point, there is current in that circuit. A perfect insulator would not allow electrons to flow through it (there would be zero current), while a perfect conductor would allow infinite current (unlimited electron flow). In practice, however, there is no such thing as a perfect conductor nor a perfect insulator. Partial opposition to electron flow occurs when the electrons collide with other electrons or atoms in the conductor. The result is a reduction in current, and the conductor produces heat.

Resistors allow us to control the current in a circuit by controlling the opposition to electron flow. As they oppose the flow of electrons they dissipate electrical energy in the form of heat. The more energy a resistor dissipates, the hotter it will become. Figure 6-1A shows the schematic symbol for all nonadjustable resistors. Part B shows a common resistor. You will learn how to interpret the colored stripes on a resistor later in this chapter.

All conductors exhibit some resistance. Thick (heavy) conductors have only a little resistance. Thinner conductors have more resistance. This is just what we would expect. Think of electron flow in a conductor as being like water flow in a pipe. If we reduce the size of the pipe, less water can flow through it. Different conductor materials also have different amounts of resistance. Copper is a good conductor, and so has a small resistance. A nickel wire has more resistance than a copper wire, if both have the same diameter.

We use numbers in the *American Wire Gauge (AWG)* scale to measure wire sizes. Smaller numbers in this scale indicate larger diameter wires. A number 12 wire

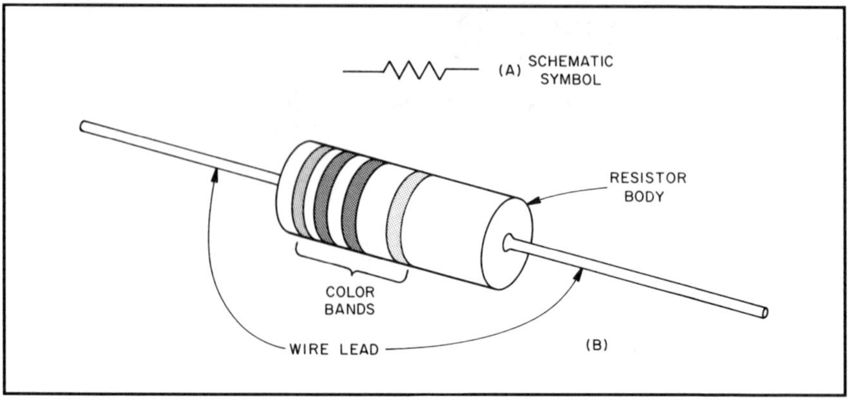

Figure 6-1—Part A shows the schematic symbol for a resistor. Part B shows a typical resistor, including the colored stripes that identify the resistor value. Resistors made in this way range in power rating from ⅛ watt to 2 watts.

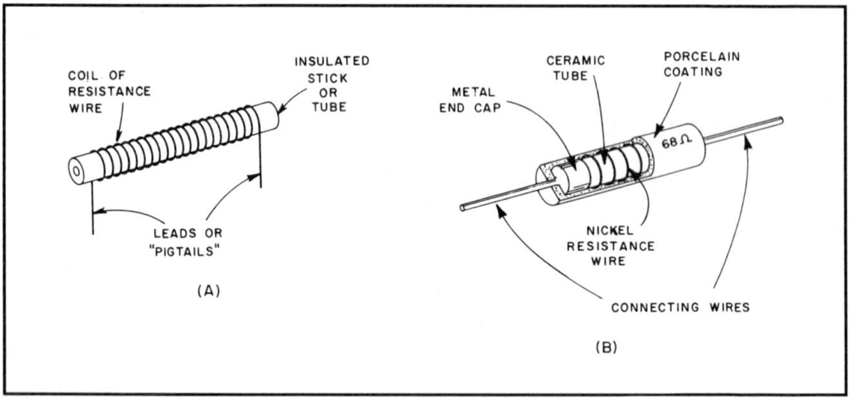

Figure 6-2—A simple wire-wound resistor is shown at A. At B is a commercially produced wire-wound resistor. These resistors are produced by winding resistive wire around a nonconductive form. Most wire-wound resistors are then covered with a protective coating.

(a common size for house wiring) has a larger diameter than a number 30 wire (which compares to the thickness of a hair).

One simple way of producing a resistor is to use a length of wire. For example, the resistance of 1000 feet of number 28 nickel wire is 337 ohms. Therefore, if we need a resistor of 3.4 ohms, we can use 10 feet of this size nickel wire. You might think 10 feet of wire makes a pretty inconvenient resistor, but we can wind it over a form of some sort. See Figure 6-2A. This makes a unit of a much more convenient size. This is precisely how **wire-wound resistors** are constructed. See Figure 6-2B.

There is a problem with this type of construction. At radio frequencies, the wire-wound construction causes the resistor to act as an **inductor**. (You will learn more about inductors later in this chapter.) It does not behave as a pure resistance. For RF circuits, resistors must be noninductive. Wire-wound resistors are common in dc circuits that carry large currents, because they can get rid of large amounts of heat.

There is another way to form a resistor. This is to connect leads to both ends of a block or cylinder of material that has a high resistance. Figure 6-3 shows a resistor made in this way. It uses a mixture of carbon and clay as the resistive element. This type of resistor is a *carbon-composition resistor*. The proportions of carbon and clay determine the value of the resistor.

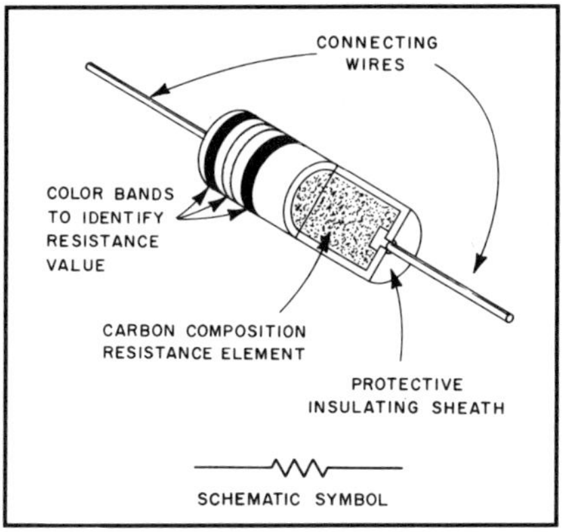

Figure 6-3—Composition resistors are made from a mixture of carbon and clay. The resistance is controlled by varying the amount of carbon in the material when the resistor is manufactured.

An advantage of carbon-composition resistors is the wide range of available values. Other advantages are low inductance and capacitance plus good surge-handling capability. The ability to withstand small power overloads without being completely destroyed is another advantage. The main disadvantage is that the resistance of the composition resistor will vary widely. Variations are caused by operating temperature changes and resistance changes as the resistor gets older.

There are two types of **film resistors**. These are made by depositing a thin layer of resistive material on a ceramic form. The ceramic form is usually shaped as a cylinder. Metal end caps provide a connection between the resistive film and the resistor leads used to connect the part in a circuit. The entire assembly then gets a protective epoxy or plastic coating.

You can easily recognize the "dog bone" shape of film resistors. They are thinner in the middle than at the ends. This is because of the metal end caps used to connect the leads to the resistive element.

Carbon-film resistors are manufactured by using high temperatures to break down certain gaseous hydrocarbons. The resulting carbon is then deposited in a thin layer or film on the ceramic form. The thickness of the deposited film provides a means to control the final resistance.

The major advantages of carbon-film resistors are low cost and improved stability with age and temperature changes. These resistors cannot withstand electrical overloads or surges. They can be used as fuses for some applications.

The *metal-film resistor* has replaced the carbon-composition type in many low-power applications today. Metal-film resistors are formed by depositing a thin layer of resistive alloy on a cylindrical ceramic form. Nichrome (an alloy made of nickel and chromium) or other materials are used. See Figure 6-4. This film is then trimmed away in a spiral fashion to form the resistance path. The trimming can be done on a mechanical lathe or by using a laser. The resistor is then covered with an insulating material to protect it.

Because of the spiral shape of the resistance path, metal-film resistors will exhibit some inductance. The effects of this inductance increase at higher frequencies. The higher the resistance of the unit, the greater the inductance will be. This is caused by the increased path length. Metal-film resistors provide much better temperature

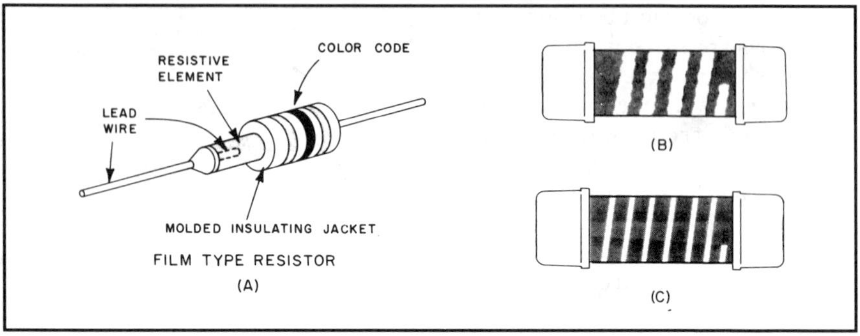

Figure 6-4—The construction of a typical metal-film resistor is shown at A. Film resistors are formed by depositing a thin film of material on a ceramic form. Material is then trimmed off in a spiral to produce the specified resistance for that particular resistor. The excess material may either be trimmed with a mechanical lathe or a laser. A lathe produces a rather rough spiral, like that shown in part B. The laser produces a finer cut, shown at C.

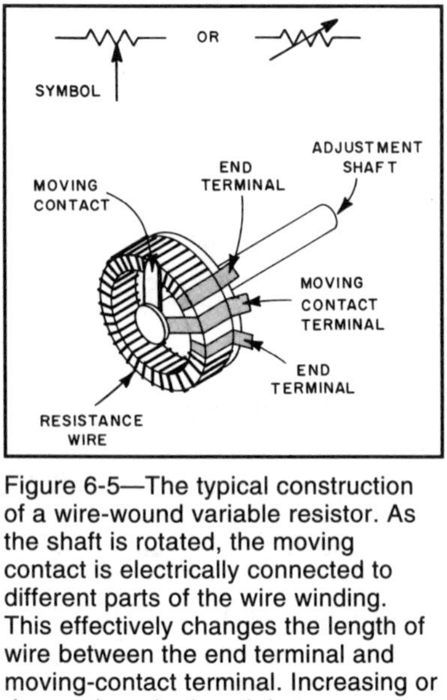

Figure 6-5—The typical construction of a wire-wound variable resistor. As the shaft is rotated, the moving contact is electrically connected to different parts of the wire winding. This effectively changes the length of wire between the end terminal and moving-contact terminal. Increasing or decreasing wire length between terminals thus increases or decreases resistance.

stability than other types.

Most resistors have standard fixed values. They are called *fixed resistors*. **Variable resistors** are sometimes called **potentiometers**. They can be used to adjust the voltage, or potential, in a circuit. Potentiometers are also used a great deal in electronics. The construction of a wire-wound variable resistor and the schematic symbol for all variable resistors is shown in Figure 6-5. Variable resistors are also made with a ring of carbon compound in place of the wire windings. A connection is made to each end of the resistance ring. A third contact is attached to a movable arm, or wiper. The wiper can be moved across the ring. As the wiper moves from one end of the ring to the other, the resistance varies from minimum to maximum.

Color Codes

It is not always practical to print the resistance values on the side of a small resistor. A **color code** shows the value of the resistor. As shown in Figure 6-6, the color of the first three bands shows the resistor value. The color of the fourth band indicates the resistor tolerance.

Standard fixed resistors are usually found in values ranging from 2.7 Ω to 22 MΩ (22 megohms, or 22,000,000 ohms). Resistance tolerances on these standard values can be ±20%, ±10%, ±5% and ±1%. A tolerance of 10% on a 200-Ω resistor means that the actual resistance of a particular unit may be anywhere from 180 Ω to 220 Ω. (Ten percent of 200 is 20, so the resistance can be 200 + 20 or 200 – 20 ohms.)

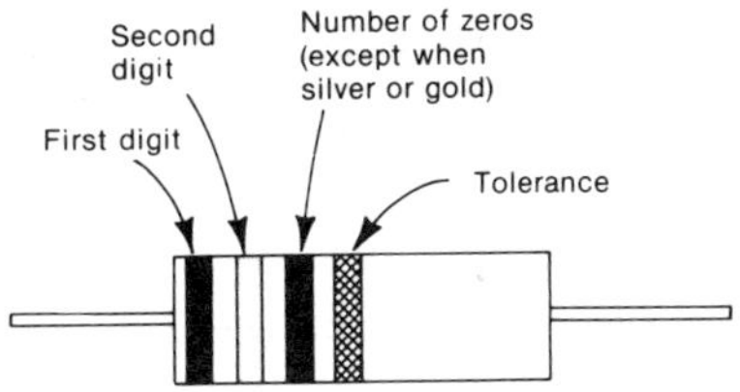

First Three Bands				Fourth Band	
Black	– 0	Blue	– 6	Gold	± 5%
Brown	– 1	Violet	– 7	Silver	± 10%
Red	– 2	Grey	– 8	none	± 20%
Orange	– 3	White	– 9		
Yellow	– 4	Silver	0.01		
Green	– 5	Gold	0.1		

Figure 6-6—Small resistors are labeled with a color code to show their value. For example, proceeding from left to right, a resistor with color bands of yellow/violet/brown/gold is a 470-W resistor with a 5% tolerance.

Here is another example to help you understand this calculation. Find the range of possible values for a 100-Ω resistor that has a 10% tolerance rating.

First calculate the range of possible variation:

$100\ \Omega \times 10\% = 100\ \Omega \times 0.10 = 10\ \Omega$

Now use this value to calculate the possible range of resistances:

$100\ \Omega - 10\ \Omega = 90\ \Omega$

$100\ \Omega + 10\ \Omega = 110\ \Omega$

A 100-Ω resistor that has a 10% tolerance rating will have a value somewhere between 90 and 110 Ω.

Suppose you are building a sensitive circuit that requires a resistor that is no more than 105 Ω. Would you select a 10%, 100-Ω resistor for this circuit? Probably not, because when you go to your local electronics parts store to buy the resistor, you may get a unit that has a resistance as high as 110 Ω. What tolerance component should you choose? A 5% tolerance resistor will suit your purpose in this circuit. You might also select a 1% resistor, because the value will only vary between 99 and 101 ohms for those parts. Resistors with the smaller tolerance rating are more expensive, however. As a general rule you should select the highest tolerance rating that will work properly in your circuit. (Higher tolerance ratings mean there is a wider range of possible values for a standard stock part.)

You can also buy resistors with even smaller tolerance ratings. *Precision resistors* are specified with 0.5%, 0.25%, 0.1% and 0.05% tolerance ratings. For a price, you could probably buy resistors with even smaller tolerances. If you have an ohmmeter that is accurate enough, you also can measure the values of standard parts to select those within the limits you require.

The tolerance rating of a resistor is not necessarily an indication of its manufactured quality. Questions T6A10 and T6A11 refer to the tolerance ratings of high-quality and low-quality resistors, however. The intent of these questions is to refer to a resistor with a small tolerance as a high-quality resistor and one with a larger tolerance as a low-quality resistor.

Power Ratings

Resistors are also rated according to the amount of power they can safely handle. Power ratings for wire-wound resistors typically start at 1 watt and range to 10 W or larger. Resistors with higher power ratings are also physically large. Greater surface area is required to dissipate the increased heat generated by large currents. Carbon-composition and film resistors are low-power components. You will find these resistors in ⅛-W, ¼-W, ½-W, 1-W and 2-W power ratings. A resistor that is being pushed to the limit of its power rating will feel very hot to the touch. Sometimes a resistor gets very hot during normal operation. You should probably replace it with a unit having a higher power-dissipation rating.

[Now turn to Chapter 10 and study all of the questions in the T6A series. This includes questions T6A01 through T6A11. Review this section as needed.]

INDUCTORS

In Chapter 5 we learned that an **inductor** stores energy in a *magnetic field*. When a voltage is first applied to an inductor, with no current flowing through the circuit, the inductor will oppose the current. Remember from Chapter 5 that inductors oppose any change in current. A voltage that opposes the applied voltage is induced in the coil, and this voltage tries to prevent a current. Gradually, though, the current will build up. Only the resistance in the circuit will limit the final current value. The resistance in the wire of the coil will be very small.

In the process of getting this current to flow, the coil stores energy. The energy is in the form of a magnetic field around the wire. The increasing field strength produces a voltage that opposes the current. When you shut off the applied current, the magnetic field collapses. The collapsing field returns the stored energy to the circuit. The energy will have the form of a momentary current in the same direction as the original current. This current is produced by the EMF generated by the collapsing magnetic field. Again, this voltage opposes any change in the current already flowing in the coil.

We can summarize the actions of an inductor by considering two factors. Inductors store energy in magnetic fields, and they oppose any change in current through the inductor. The answer to question T6B02 refers to storing a charge electromagnetically. Electromagnetic here means an *electric current* produces the *magnetic field*. An inductor **does not** store electric *charge*, however. It stores *energy*.

The amount of opposition of an inductor to changes in current is called *inductive* **reactance**. The reactance of the inductor, the amount of energy stored in the magnetic field and the back EMF induced in the coil all depend on the amount of inductance.

We use a capital L to represent the amount of inductance that a coil exhibits. Inductance depends on four things:

1) the type of material used in the core, and its size and location in the coil
2) the coil diameter
3) the length of the coil and
4) the number of turns of wire used to wind the coil

Changing any of these factors changes the inductance.

An inductor's **core** refers to the area inside the inductor, where the magnetic field is concentrated. If we add an *iron* or *ferrite core* to the coil, the inductance increases. We can also use brass as a core material in radio-frequency variable inductors. It has the opposite effect of iron or ferrite: The inductance decreases as the brass core enters the coil. A movable core can be placed in a coil, as shown in Figure 6-7. The amount of inductance can be varied as the core moves in and out of the coil. The common schematic symbols for various inductors are shown in Figure 6-8.

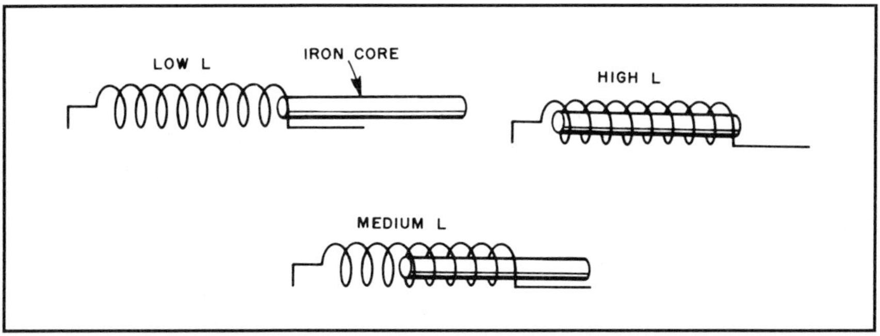

Figure 6-7—The inductance of a coil increases as an iron core is inserted. Many practical inductors of this type use screw threads to allow moving the core in precise amounts when making small changes in inductance.

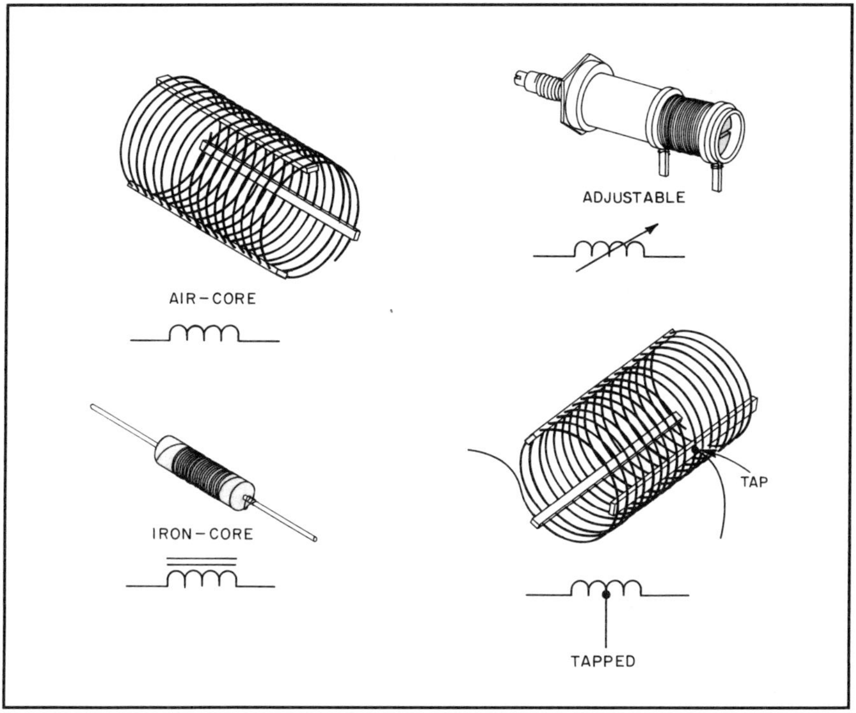

Figure 6-8—Various types of coils, and their schematic symbols. The adjustable inductor uses a movable powdered-iron core or slug. They are sometimes called slug-tuned or permeability-tuned inductors.

The adjustable coil shown in Figure 6-8 has a threaded core so you can adjust it into and out of the coil. Such inductors usually serve to adjust the operating frequency of a circuit. These adjustments often require you to turn the core material into or out of the coil while the circuit is operating. You must use a plastic tool to make these adjustments rather than a metal tool like a screwdriver. Bringing a metal tool close to the coil would change the inductance. This would result in an incorrect circuit adjustment

Ferrite comes from the Latin word for iron. So what is the difference between an iron and a ferrite core? Well, *iron cores* can be made from iron sheet metal. The layers of material that make up an ac power transformer core are made that way. Cores can also be made from powdered iron mixed with a bonding material. This holds the powdered iron together so it can be molded into the desired shape. Manufacturers mix other metal alloys with the iron "ferrous" material when they make some cores. We call these *ferrite cores*. They can select the proper alloy to provide the desired characteristics, producing cores designed to operate best over a specific frequency range. We use iron cores most often at low frequencies, such as audio frequencies. Audio and ac-power circuits are good applications for these cores. Ferrite cores are manufactured in a wide range of compositions. This optimizes their operation for specific radio-frequency ranges.

It takes some amount of energy to magnetize the core material for any inductor. This energy represents a loss in the coil. Winding the coil on an iron material increases the inductance of the coil. The energy needed to magnetize the core also increases. Some coils are wound on thin plastic forms, or made self supporting. Then the only material inside the coil is air. These are air-wound coils. They are among the lowest loss types of inductors. Air-wound coils are normally used in high-power RF circuits where energy loss must be kept to a minimum.

Toroid Cores

A *toroidal inductor* is made by winding a coil of wire on a donut-shaped core called a toroid. **Toroidal core** materials are usually powdered-iron or ferrite compounds. The toroidal inductor is highly efficient. This is because there is no break in the circular core. All the magnetic lines of force remain inside the core. This means there is very little *mutual coupling* between two toroids mounted close to each other in a circuit.

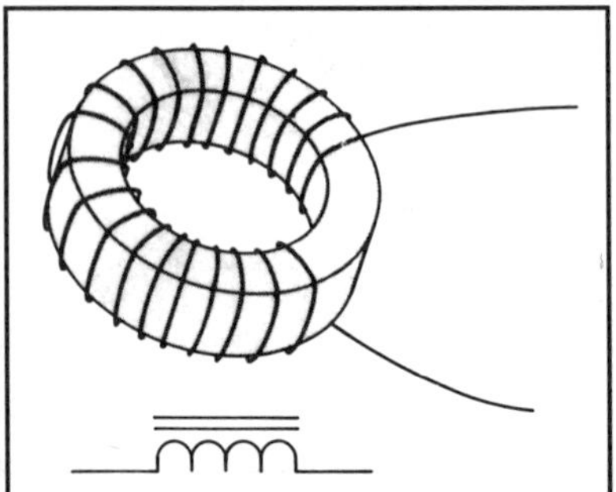

Figure 6-9—A toroidal coil. This type of coil has self-shielding properties. Some inductors in some circuits will interact with the magnetic fields of other inductors. To prevent this a metal shield is used to enclose them. Toroidal inductors confine magnetic fields so well that shields are usually not necessary.

Figure 6-9 shows a toroidal inductor and the schematic symbol used to represent them. Notice that the schematic symbol is the same one used for any inductor that has an iron or ferrite core material.

Mutual coupling occurs when the magnetic flux of one coil passes through the windings of the other coil. When coils have mutual coupling, a current in one coil will induce a voltage (and a current) in the other. Mutual coupling can be a problem, especially in an RF circuit. There, signals from one stage could be coupled into another stage without following the proper signal path. Toroidal cores can often help eliminate the mutual coupling that would occur between other types of coils.

In theory, a toroidal inductor requires no shield to prevent its magnetic field from spreading out and interfering with outside circuits. This self-shielding property also helps keep outside forces from interfering with the toroidal coil. Toroids can be mounted so close to each other that they can almost touch, but because of the way they are made, there will be almost no inductive coupling between them. Toroidal coils wound with only a small amount of wire on ferrite or iron cores can have a very high inductance value.

[Stop here and turn to Chapter 10 now. Study exam questions T6B01 through T6B08. Review any material you have difficulty with before going on.]

CAPACITORS

A **capacitor** is composed of two or more conductive plates with an insulating material between them. In Chapter 5 we learned that the basic property of a capacitor is the ability to store an *electric charge*. In an uncharged capacitor, the potential difference between the two plates is zero. As we charge the capacitor, this potential difference increases until it reaches the full applied voltage. The difference in potential creates an **electric field** between the two plates. See Figure 6-10.

A field is an invisible force of nature. We put energy into the capacitor by charging it. Until we discharge it, or the charge leaks away somehow, the energy is

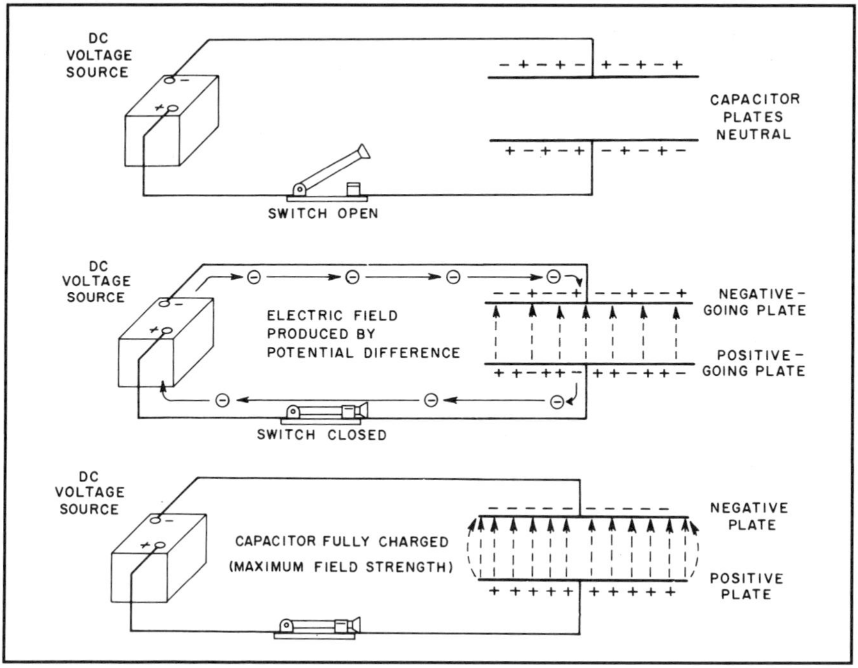

Figure 6-10—Capacitors store energy in an electric field. The final amount of energy stored in a fully charged capacitor depends on the capacitor value and the applied voltage. If this circuit is used to fully charge the capacitor from a 6-volt battery, the amount of energy stored in the capacitor could be increased by using a 12-volt battery. Another way to increase the amount of stored energy is to use a larger-value capacitor.

stored in the electric field. When the field is not moving, we sometimes call it an *electrostatic field.*

When we apply a dc voltage to a capacitor, current will flow in the circuit until the capacitor is fully charged to the applied voltage. After the capacitor has charged to the full applied voltage, no more current will flow in the circuit. For this reason, capacitors can be used to block dc in a circuit. Because ac voltages are constantly reversing polarity, we can block dc with a capacitor while permitting ac to pass.

When the voltage across a capacitor increases, the capacitor stores more charge in reaction to the increased voltage. When the voltage decreases, the capacitor returns some of its stored charge. These actions tend to oppose any change in the voltage applied to a capacitor. This opposition to changes in voltage (and current) is called *capacitive* **reactance**.

We can summarize the operation of a capacitor by considering two factors. A capacitor stores electric charge in the form of an electrostatic field and it opposes any change in the applied voltage.

Factors That Determine Capacitance

Remember from Chapter 5 that the value of a capacitor is determined by three factors:

1) the type of insulating material or dielectric used between the plates
2) the area of the plate surfaces
3) the distance between the plates

There are two factors involved with the surface area of the plates. The first factor is the number of plates. By adding additional sets of plates, each separated by insulating material, you can increase the effective plate area. Figure 6-11 shows the construction of a capacitor with two plates for each side. Additional layers of foil and insulating material can be added to increase the capacitance. Alternate foil layers connect to opposite sides of the capacitor in this construction.

The second surface-area factor involves knowing the surface area of one side of one plate. If you measure this surface area, and know how many plates are used to build a certain capacitor, you could calculate its capacitance. (You would also have to know the *dielectric constant* of the insulator and the distance between the plates.)

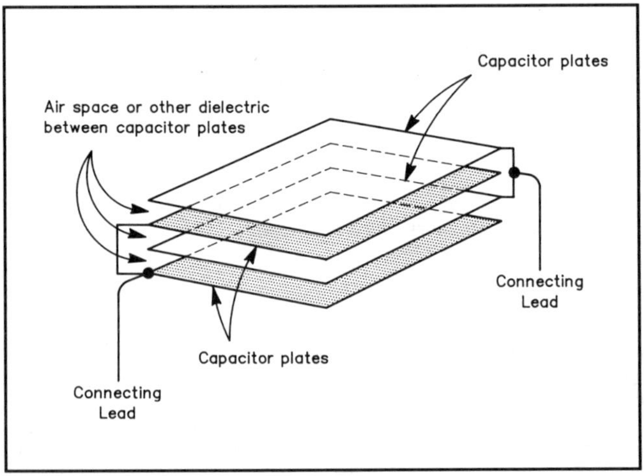

Figure 6-11—This drawing shows how the total plate surface area can be increased by stacking several plates, with insulating material between them. This increases the capacitance of the unit, compared to the capacitance if only two plates of the same size are used. The leads connect to alternate plates.

As the surface area increases, the capacitance increases. For a given plate area, as the spacing between the plates decreases, the capacitance increases.

Practical Capacitors

Practical capacitors are described by the material used for their dielectric. Mica, ceramic, plastic-film, polystyrene, paper and electrolytic capacitors are in common use today. They each have properties that make them more or less suitable for a particular application. Figure 6-12A shows the symbol used to represent capacitors on schematic diagrams.

Electrolytic capacitors have a dielectric that is formed after the capacitor is manufactured. Two common types of electrolytic capacitors are aluminum-electrolytic capacitors and tantalum-dielectric capacitors.

Electrolytic capacitors can be made with a very high capacitance value in a small package. Electro-

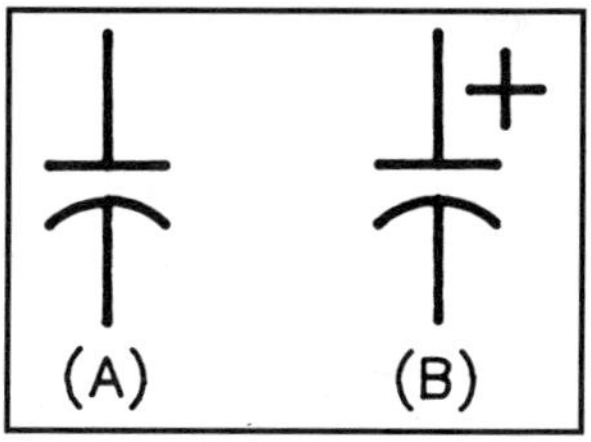

Figure 6-12—A shows the schematic symbol representing fixed capacitors. This symbol has the appearance of two plates separated by an insulator. B shows the symbol representing electrolytic capacitors, which must be connected to dc voltages with the proper polarity.

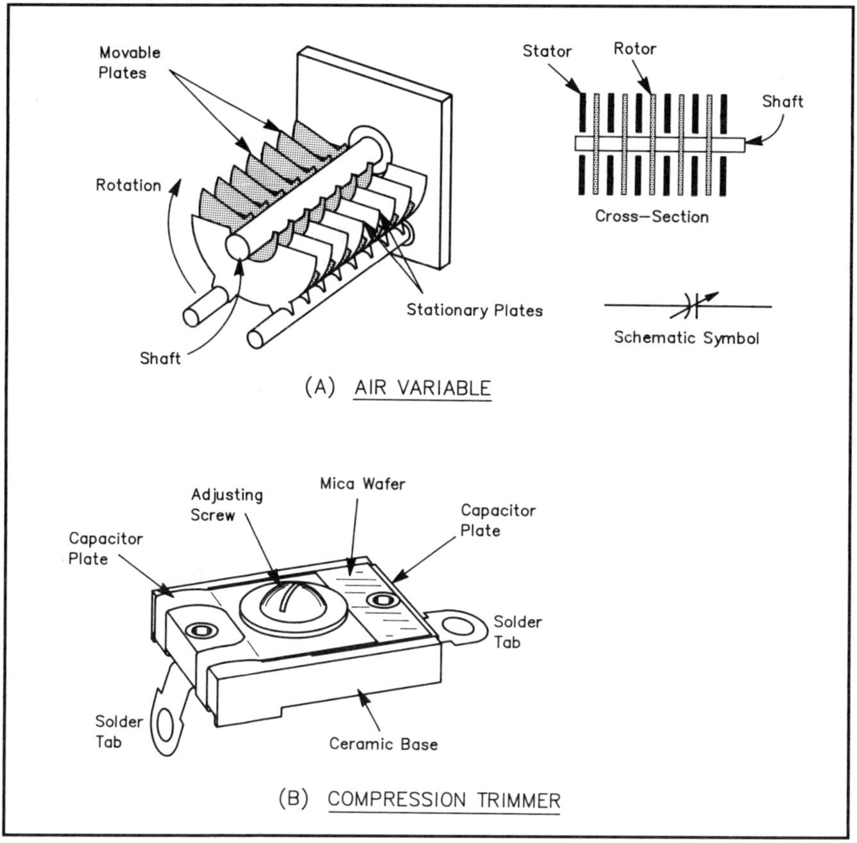

Figure 6-13—Variable capacitors can be made with air as the dielectric, or with mica or ceramic dielectrics.

lytic capacitors are polarized — dc voltages must be connected to the positive and negative capacitor terminals with the correct polarity. The positive and negative electrodes in an electrolytic capacitor are clearly marked. Connecting an electrolytic capacitor incorrectly causes gas to form inside the capacitor and the capacitor may actually explode. This can be very dangerous. At the very least the capacitor will be destroyed by connecting it incorrectly.

When we draw electrolytic capacitors on schematic diagrams, we must indicate the proper polarity for the capacitor. We do this by adding a + sign to the capacitor symbol, as Figure 6-12B shows.

In some circuits it is necessary or desirable to vary the capacitance at some point. An example is in the tuning circuit of a receiver or transmitter VFO. We could use a rotary switch to select one of several different fixed-value capacitors. It is much more convenient to use a **variable capacitor**, however. A basic air-dielectric variable capacitor is shown in Figure 6-13A. This figure also shows the schematic symbol for a variable capacitor. One set of plates is fixed in one position to the capacitor frame. These plates form the *stator*. The other set of plates rotates by turning the control shaft. The movable plates form the *rotor*. The capacitance can be varied smoothly between the maximum and minimum values.

Another type of variable capacitor is the compression variable shown in Figure 6-13B. In this type of variable capacitor, the spacing between the two plates is varied. This type of variable capacitor is usually used as a *trimmer capacitor*. A trimmer capacitor is a control used to peak or fine tune a part of a circuit. It is usually left alone once adjusted to the correct value. You should use a plastic screwdriver-like adjusting tool when tuning a trimmer capacitor. The metal of a normal screwdriver can affect the capacitance and cause false readings as you try to adjust it.

[This completes your study of Chapter 6. Now turn to Chapter 10 and study questions T6B09 through T6B14. Review any material you have difficulty with before going on to Chapter 7.]

KEY WORDS

Amplifier—A device usually employing electron tubes or transistors to increase the voltage, current or power of a signal. The amplifying device may use a small signal to control voltage and/or current from an external supply. A larger replica of the small input signal appears at the device output.

Band-pass filter—A circuit that allows signals to go through it only if they are within a certain range of frequencies. It attenuates signals above and below this range.

Beat-frequency oscillator (BFO)—An oscillator that provides a signal to the product detector. In the product detector, the BFO signal and the IF signal are mixed to produce an audio signal.

Detector—The stage in a receiver in which the modulation (voice or other information) is recovered from the RF signal.

Filter—A circuit that will allow some signals to pass through it but will greatly reduce the strength of others.

Frequency discriminator—A circuit that produces a waveform which varies in amplitude as the incoming signal frequency changes. A discriminator is the **detector** in an FM receiver.

Frequency modulation (FM)—The process of varying the frequency of an RF carrier in response to the instantaneous changes in the modulating signal.

High-pass filter—A filter that allows signals above the cutoff frequency to pass through. It attenuates signals below the cutoff frequency.

Intermediate frequency (IF)—The output frequency of a mixing stage in a superheterodyne receiver. The subsequent stages in the receiver are tuned for maximum efficiency at the IF.

Limiter—An FM-receiver stage between the IF amplifier and the **frequency discriminator**, which chops noise and amplitude variations from an incoming signal. The limiter produces a constant-amplitude signal for the discriminator.

Low-pass filter—A filter that allows signals below the cutoff frequency to pass through and attenuates signals above the cutoff frequency.

Modulate—To vary the amplitude, frequency, or phase of a radio-frequency signal.

Modulation—The process of varying a radio wave in some way to send information.

Phase modulation (PM)—Varying the phase of an RF carrier in response to the instantaneous changes in the modulating signal.

Radio-frequency (RF) carrier—A constant-amplitude, unmodulated, radio-frequency signal.

Reactance modulator—A device capable of modulating an ac signal by varying the reactance of a circuit in response to the modulating signal. (The modulating signal may be voice, data, video, or some other kind depending on what type of information is being transmitted.)

Variable-frequency oscillator (VFO)—An oscillator used in receivers and transmitters. The frequency is set by a tuned circuit using capacitors and inductors. The frequency can be changed by adjusting the components in the tuned circuit.

7 Practical Circuits

In this chapter we will discuss low-pass, high-pass and band-pass filter circuits. We'll show you block diagrams of complete transmitters and receivers, and investigate how the stages connect to make them work. After each small bite of text, we'll tell you to turn to Chapter 10 and study some of the questions in the Technician Question Pool. The questions will help you review your understanding of the text material. If you have any difficulty, review those sections of this chapter.

Keep in mind that entire books have been written on each topic covered in this section. You may not understand some of the circuits from our brief discussion. It would be a good idea to consult some other reference books. ARRL's *Understanding Basic Electronics* is a good starting point. *The ARRL Handbook for Radio Amateurs* includes plenty of detail about these topics. Even that book won't tell you everything about a topic, though. The discussion in this chapter will help you understand the circuits well enough to pass your Technician class license exam, however.

FILTERS

Harmonic interference to entertainment equipment is a problem some hams face. Harmonics are multiples of a given frequency. Your transmitter radiates undesired harmonics along with your signal. In the high-frequency amateur bands (3.5 to 29.7 MHz), the frequency you're transmitting on is much lower than the TV or FM channels. Some of your harmonics may fall within the home entertainment bands. The entertainment receiver cannot distinguish between the TV or FM signals that are supposed to be there and your harmonics, which are not. If your harmonics are strong enough, they can interfere with the broadcast signal. Modern Amateur Radio transceivers have filters built in to *attenuate* (reduce) harmonic radiation so they seldom create harmonic-related interference.

Harmonic interference must be cured at your transmitter. It is your responsibility as a licensed amateur. You must see that harmonics from your transmitter are not strong enough to interfere with other services. As mentioned before, all harmonics generated by your transmitter must be attenuated well below the strength of the fundamental frequency. If harmonics from your transmitting equipment exceed these limits, you are at fault.

You can usually tell harmonic interference when you see it on a TV set. This type of interference shows up as crosshatch or a herringbone pattern on the TV screen. See Figure 7-1. Unlike overload, interference from radiated harmonics seldom affects all channels. Rather, it may bother the one channel that is frequency-related to the band you're on. Generally, the low TV channels (channels 2-6) are most affected by harmonics from amateur transmitters operating below 30 MHz. Channels 2 and 6 are

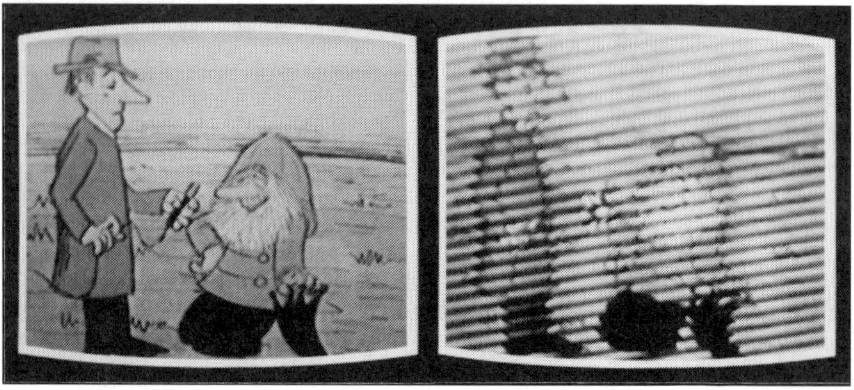

Figure 7-1—Harmonic interference may be visible as cross-hatching of the TV screen. More severe interference can completely destroy the picture. The width of the cross-hatch lines will vary, depending on the transmitter frequency.

especially affected by 10-meter transmitters, and channels 3 and 6 experience trouble from 15-meter transmitters.

There are several possible cures for harmonic interference. We will discuss a few of them here. Try each step in the order we introduce them. Chances are good that your problem will be quickly solved. You should be familiar with three **filters**: the **low-pass filter**, the **high-pass filter** and the **band-pass filter**.

Filters pass certain frequencies and block others. All modern radio communication devices use filter circuits. These filter circuits allow various kinds of equipment to operate on different frequencies without interfering with each other. Under certain conditions, some equipment may need a little extra help. A transmitter may need extra filtering of the output signal to reduce interference to television receivers. In other cases, the transmitter may already be well filtered (clean), but a TV receiver may need extra filtering of its input signals. A basic understanding of filters and their applications will often allow you to solve interference problems.

Low-Pass Filters

The first step you should take is installation of a **low-pass filter** in the transmission line between your transmitter and antenna. A low-pass filter is one that passes all frequencies below a certain frequency, called the *cutoff frequency*. We measure the filter cutoff frequency by putting a variable-frequency signal into the filter. The input signal power is kept constant while the frequency is increased. At the same time we measure the filter output power. At some frequency the output begins to decrease. When the output power has decreased to ½ the input power, we have found the cutoff frequency. Frequencies above the cutoff frequency are *attenuated,* or significantly reduced in amplitude. See Figure 7-2.

The cutoff frequency depends on the design of the low-pass filter. The capacitors are chosen to provide a path to ground for the unwanted high frequencies. The inductor has a value that passes the low frequencies you want, but blocks the higher frequencies.

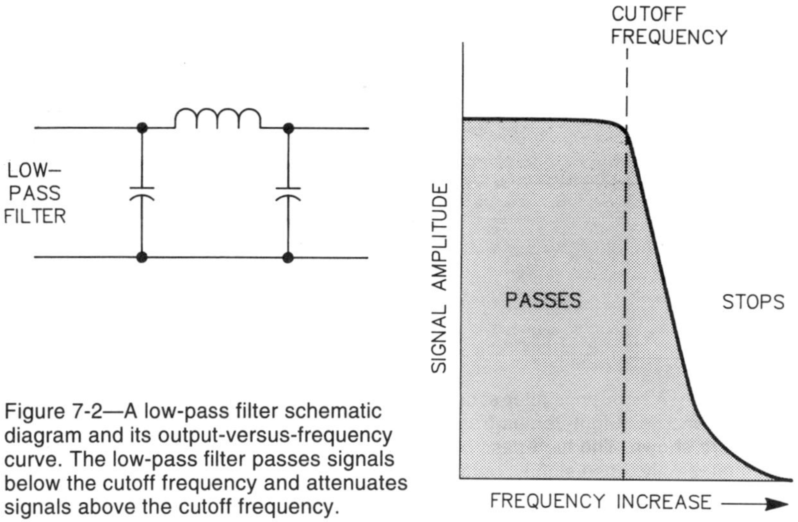

Figure 7-2—A low-pass filter schematic diagram and its output-versus-frequency curve. The low-pass filter passes signals below the cutoff frequency and attenuates signals above the cutoff frequency.

One other important trait that must be considered in the design of a low-pass filter is the *characteristic impedance*. This is the circuit impedance for which the filter is designed. Most external low-pass filters intended for amateur use will have a characteristic impedance of 50 ohms.

The low-pass filter is always connected between the transmitter and the antenna, as close to the transmitter as possible. The filter should have the same impedance as the feed line connecting the transmitter to the antenna. Filters should be used only in a feed line with a low standing-wave ratio. The cutoff frequency has to be higher than the highest frequency used for transmitting. A filter with a 45-MHz cutoff frequency would be fine for the high-frequency (HF) bands. This filter would significantly attenuate 6-meter (50-MHz) signals, so it should not be used for 50 MHz or higher-frequency operation. Most modern transmitters have a low-pass filter built into their output circuitry to prevent excess harmonic radiation.

High-Pass Filters

Sometimes entertainment devices experience interference from amateur transmissions even though the transmitting device is operating properly. The harmonics can be well below levels necessary to prevent interference. This happens because the design of the entertainment device is inadequate when it is operating in the presence of strong signals. In this situation, it may be necessary to reduce the level of the amateur signal reaching the entertainment device (TV or FM stereo receiver, or VCR). The desired higher-frequency signals must be allowed to pass unaffected. A **high-pass filter** can do this.

A high-pass filter passes all frequencies above the *cutoff frequency*, and *attenuates* those below it. See Figure 7-3. A high-pass filter should be connected to a

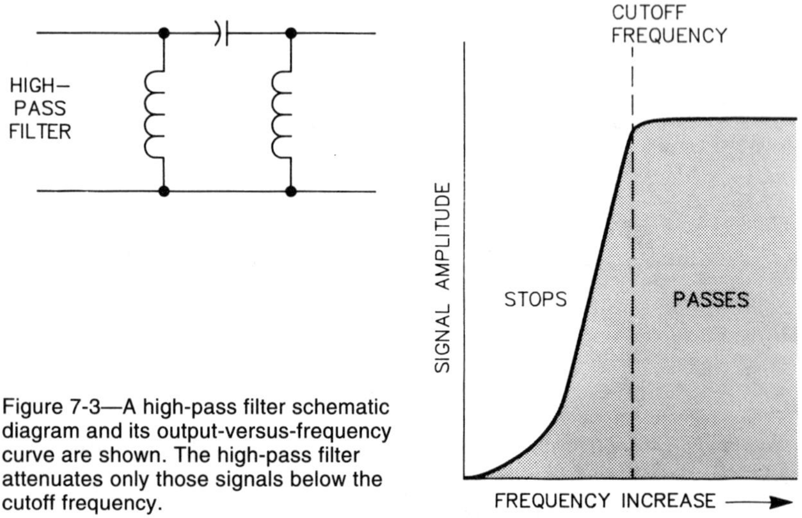

Figure 7-3—A high-pass filter schematic diagram and its output-versus-frequency curve are shown. The high-pass filter attenuates only those signals below the cutoff frequency.

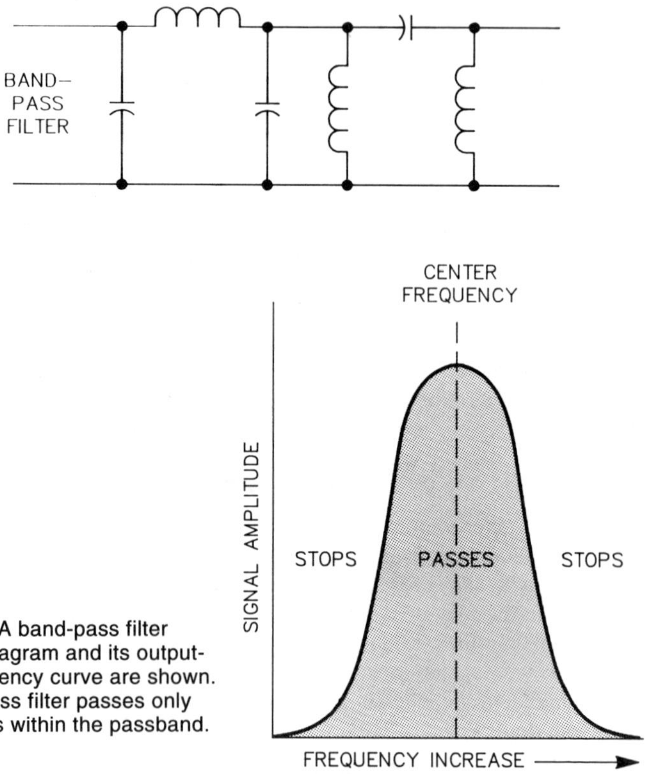

Figure 7-4—A band-pass filter schematic diagram and its output-versus-frequency curve are shown. The band-pass filter passes only those signals within the passband.

television set, stereo receiver or other home-entertainment device that is being interfered with. It will attenuate the signal from an amateur station. The inductors have a value that allows them to conduct the lower frequencies to ground. The capacitor tends to block these low-frequency signals. At the same time it allows the higher-frequency television or FM-broadcast signals to pass through to the receiver. This is useful in reducing the amount of lower-frequency signal (at the ham's operating frequency). The lower-frequency signal might overload a television set, causing disruption of reception.

For best effect, a high-pass filter should be connected as close to the television-set tuner as possible. Put it inside the TV if it is your own set and you don't mind opening the cabinet. You can also attach it directly to the antenna terminals on the back of the television. If the set belongs to your neighbors, have them contact a qualified service technician to install the filter. That way you are not held responsible if something goes wrong with the set later on.

Band-Pass Filters

A **band-pass filter** is a combination of a high-pass and low-pass filter. It passes a desired range of frequencies while rejecting signals above and below the *pass band*. This is shown in Figure 7-4.

Band-pass filters are commonly used in the **intermediate-frequency (IF)** stage of a receiver to provide different degrees of rejection. Very narrow filters are used for CW reception. Wider filters are switched in for SSB and AM double-sideband reception. (SSB and AM are explained later.) The band-pass filter allows the receiver IF stages to select signals within a certain frequency range and block signals outside that range.

[Now turn to Chapter 10 and study exam questions T7A01, T7A02 and T7A03. Review this section if you have trouble with any of these questions.]

TRANSMITTERS

Amateur transmitters range from the simple to the elaborate. They generally have two basic stages, an *oscillator* and a *power amplifier (PA)*. This chapter introduces the CW transmitter and the FM transmitter.

Separate receivers and transmitters have been replaced by transceivers in most amateur stations. A transceiver combines circuits necessary for receiving and transmitting in one package. Some circuit sections may perform both transmitting and receiving functions. Other sections are dedicated to transmit-only or receive-only functions. To simplify some of the following information, we will speak of transmitter or receiver circuits. The principles of operation are the same in transceivers. The transmitter topics apply to transceivers in the transmit mode. Receiver topics apply to transceivers in the receive mode.

An *oscillator* produces an ac waveform with no input except the dc operating voltages. You may be familiar with audio oscillators, like a code-practice oscillator. An RF oscillator can be used by itself as a simple low-power transmitter, but the power output is very low. See Figure 7-5. In a practical transmitter the signal from the oscillator is usually fed to one or more **amplifier** stages.

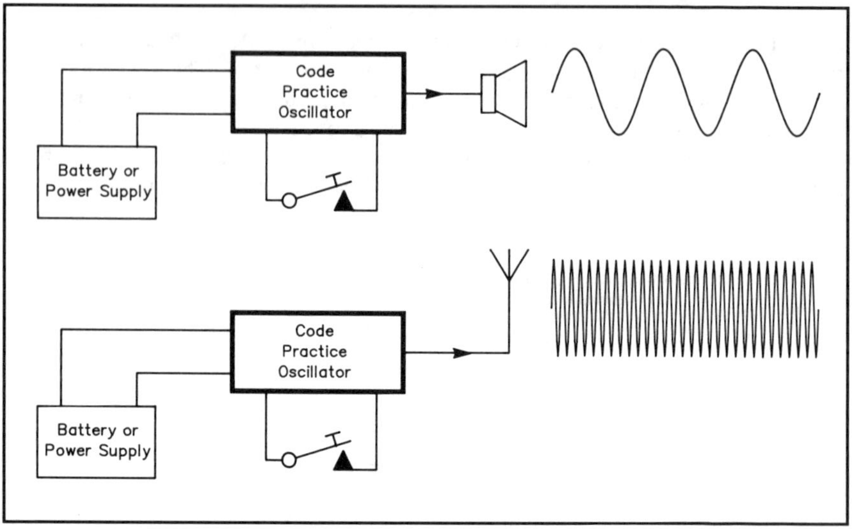

Figure 7-5—Audio- and radio-frequency oscillators are shown. Even a low-power RF oscillator connected to a good antenna can send signals hundreds of miles.

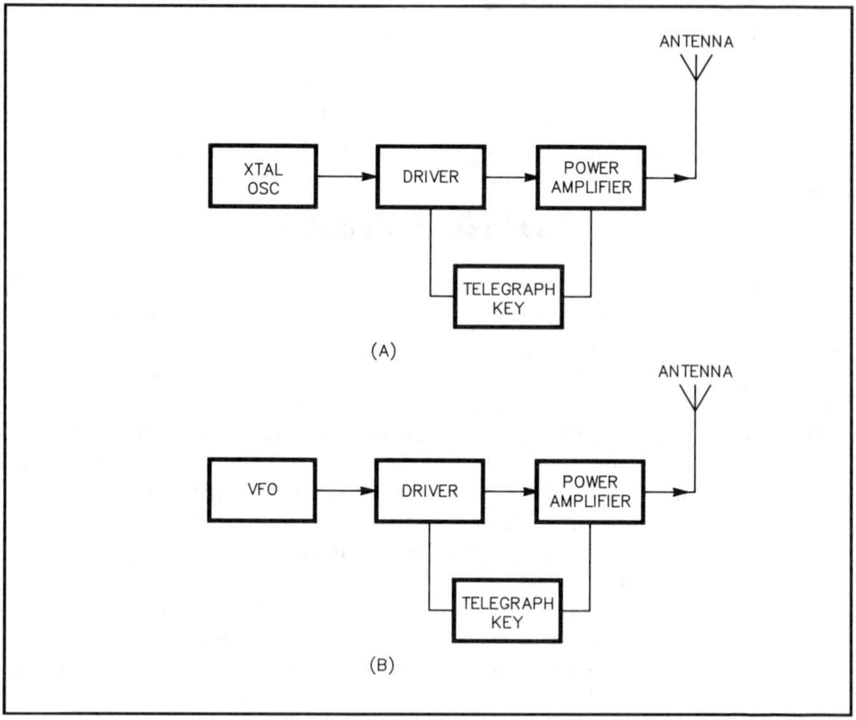

Figure 7-6—Part A shows a block diagram of a basic crystal-controlled CW transmitter. B shows a simple VFO-controlled CW transmitter.

CW Transmitters

A block diagram of a simple amateur transmitter is shown in Figure 7-6A. It produces CW (continuous-wave) radio signals when the key is closed. The simple transmitter consists of a crystal oscillator followed by a driver stage and a power amplifier. A crystal oscillator uses a quartz crystal to keep the frequency of the radio signal constant.

To transmit information, we must **modulate** the radio signal. This means we are making some change to the radio signal, and adding information to it. Pressing the key of a CW transmitter and holding it down produces a **radio-frequency (RF) carrier**. This is also called an *unmodulated* carrier. An RF carrier is a constant-amplitude, unmodulated, radio-frequency signal. When you use the key to turn the signal on and off to produce Morse code dots and dashes, however, you modulate the signal. **Modulation** is the process of varying an RF carrier in some way to send information.

Crystal oscillators may not be practical in all cases. A different crystal is needed for each operating frequency. After a while, this becomes quite expensive, as well as impractical. If we use a **variable-frequency oscillator (VFO)** in place of the crystal oscillator, as shown in Figure 7-6B, we can change the transmitter frequency whenever we want.

The telegraph key may connect to one or more transmitter sections. Figure 7-6 shows the key connected to both the driver and power amplifier stages. You would not normally key the oscillator stage because this may cause the oscillator to change frequency slightly.

All transmitters also require some source of operating voltage. We have left the power supply block off the Figure 7-6 drawings for simplicity.

Great care in design and construction is required if the stability of a VFO is to compare with that of a crystal oscillator. Stability means the ability of a transmitter to remain on one frequency without drifting. The frequency-determining components of a VFO are very susceptible to changes in temperature, supply voltage, vibration and changes in the amplifiers following the VFO. These factors have far less effect on the frequency of a vibrating quartz crystal.

FM Transmitters

When a radio signal or carrier is modulated, some characteristic of the radio signal is changed in order to convey information. We transmit information by modulating any property of a carrier. For example, we can modulate the frequency or phase of a carrier. Frequency modulation and phase modulation are closely related. The phase of a signal cannot be varied without also varying the frequency, and vice versa. **Phase modulation (PM)** and **frequency modulation (FM)** are especially suited for channelized local UHF and VHF communication. They feature good audio fidelity and high signal-to-noise ratio.

A simple FM transmitter is shown in Figure 7-7. Remember that a circuit containing capacitance and inductance is resonant at some frequency. A resonant circuit in the feedback path of an oscillator can control the oscillator frequency. By changing the resonant frequency of the tuned circuit, we can change the oscillator frequency.

A capacitor microphone is just a capacitor with one movable plate (the diaphragm). When you speak into the microphone, the diaphragm vibrates and the spacing between the two plates of the capacitor changes. When the spacing changes, the

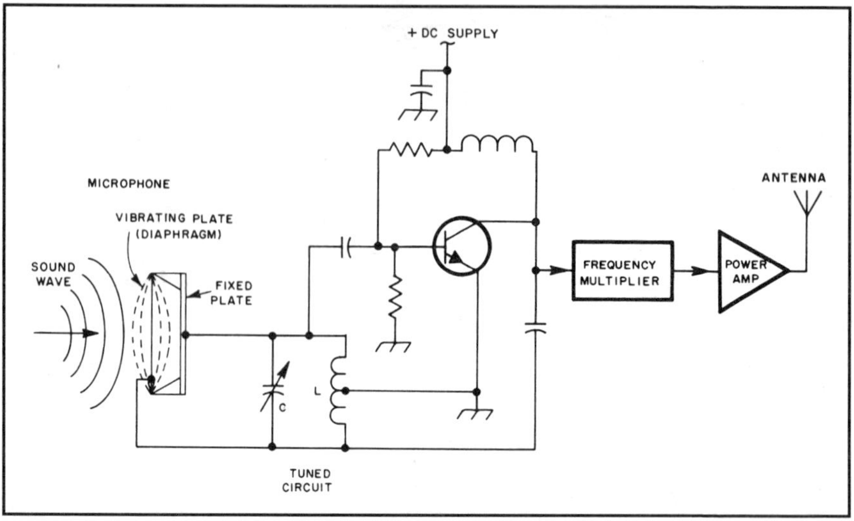

Figure 7-7—This diagram shows a simple FM transmitter. The frequency of the oscillator is changed by changing the capacitance in the resonant circuit. The vibrating plate in the microphone is part of the total capacitance in the resonant circuit.

capacitance changes. We can connect the microphone to the resonant circuit in our oscillator. Speaking into the microphone varies the oscillator frequency. Notice that if the microphone on your FM transmitter stops working, you will transmit an *unmodulated* **RF carrier**.

We can add frequency multipliers to bring the oscillator frequency up to our desired operating frequency. Adding an amplifier to increase the power provides a simple FM transmitter.

In practical FM systems, the carrier frequency is varied or modulated by changes in voltage that represent information to be transmitted. This information may originate from a microphone, a computer modem or even a video camera. The carrier frequency changes in proportion to the rise and fall of the modulating voltage.

In other words, a modulating voltage that is becoming more positive increases the carrier frequency. As the voltage becomes more negative, the carrier frequency decreases. For some purposes, reversing this relationship will work just as well (FM voice transmitters, for instance). The main point is, the carrier frequency is increased and decreased by the modulating signal voltage. The amount of increase or decrease depends on how much the voltage of the modulating signal changes. Speaking loudly into the microphone causes larger variations in carrier frequency than speaking in a normal voice. Speaking *too* loudly, though, may cause distortion.

One way to shift the frequency of the oscillator is to use a **reactance modulator**. A reactance modulator uses a vacuum tube or transistor. It is connected so that it changes either the capacitance or inductance of the oscillator resonant circuit. The changes occur in response to an input signal.

Modern FM transmitters may use a *varactor diode* to modulate the oscillator. A varactor diode changes capacitance when its bias voltage changes. You can connect a varactor diode to a crystal oscillator as shown in Figure 7-8. The varactor diode will "pull" the oscillator frequency slightly when the audio input changes.

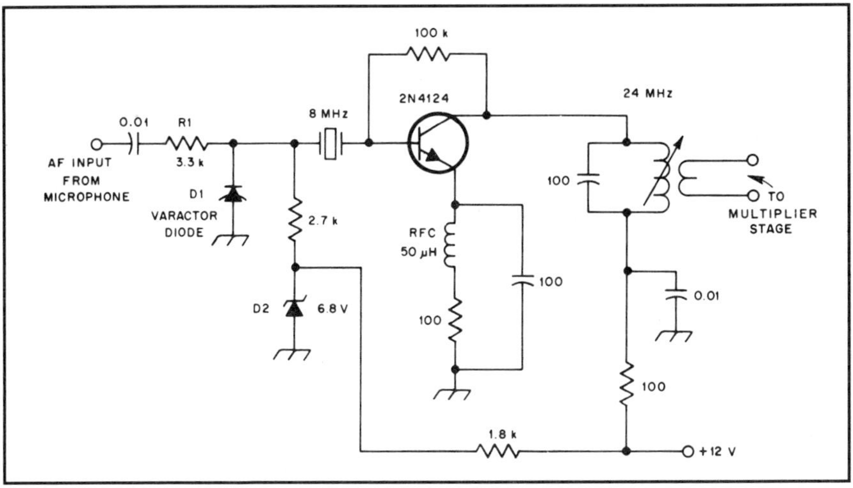

Figure 7-8—A varactor diode (D1) can be used to frequency modulate an oscillator. The capacitance of the varactor diode changes when the bias voltage varies.

Phase Modulation

One problem with direct-modulated FM transmitters is frequency stability. Designs that allow the oscillator frequency to be easily modulated may have an unfortunate side effect. It becomes more difficult to minimize unwanted frequency shifts arising from changes in temperature, supply voltage, vibration and so on. The frequency multiplying stages also multiply any drift or other instability problems in the oscillator. With phase modulation, the modulation takes place after the oscillator stage. **Phase modulation** produces what is called *indirect FM.*

The most common method of generating a phase-modulated telephony signal is to use a reactance modulator. A vacuum tube or a transistor is connected so that it changes either the capacitance or inductance of a resonant circuit in response to an input signal. The RF carrier is passed through this resonant circuit. Changes in the resonant circuit caused by the reactance modulator cause phase shifts in the RF carrier. Figure 7-9 shows the connection of a reactance modulator.

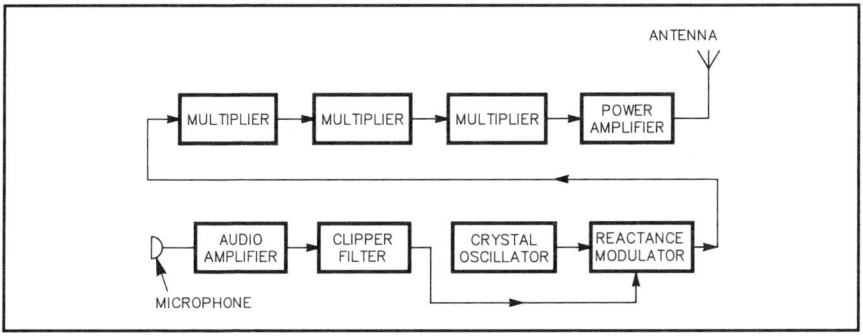

Figure 7-9—This is a block diagram of a phase-modulated transmitter, showing the reactance modulator.

[Now turn to Chapter 10 and study questions T7A05, T7A08 and T7A11. Also study questions T8A09, T8A10 and T8B01 through T8B03. Review this section as needed.]

RECEIVERS

The main purpose of any radio receiver is to change radio-frequency signals (which we can't hear or see) to signals that we can hear or see. A good receiver can *detect* weak radio signals. It separates them from other signals and interference. Also, it stays tuned to one frequency without drifting. The ability of a receiver to detect weak signals is called *sensitivity*. *Selectivity* is the ability to separate (select) a desired signal from undesired signals. *Stability* is a measure of the ability of a receiver to stay tuned to a particular frequency. In general, then, a good receiver is very sensitive, selective and stable.

Amateur receivers can be simple or complex. You can build a simple solid-state receiver that will work surprisingly well. *The ARRL Handbook for Radio Amateurs* has receiver plans, including sources for parts and circuit boards.

Detection

The **detector** is the heart of a receiver. It is where we collect the information we want from the signal. "Crystal sets" using galena crystals and "cat's whiskers" were an early form of *amplitude modulation (AM)* detector. AM is generated by varying the amplitude of an RF signal in response to a microphone or other signal source. The cat's whisker is just a thin, stiff piece of wire. With the galena crystal it forms a point-contact diode. See Figure 7-10.

Today, however, more sensitive and selective receivers are required. In addition, crystal sets cannot receive single sideband (SSB) and CW signals properly. The crystal set shows how simple a receiver can be. It is an excellent example of how detection works. Every receiver has some type of detector. Figure 7-11A shows another simple receiver. Diode D1 does the same job as the cat's-whisker detector.

Some later AM receivers had circuits similar to Figure 7-11B. This is a tuned-radio-frequency (TRF) receiver. An incoming signal causes current to flow from the antenna, through the resonant tuned circuit, to ground. The current induces a voltage with the same waveform in L2. The L1-C1 circuit resonates at the frequency of the incoming signal and tends to reject signals at other frequencies. The diode rectifies the RF signal, allowing only half the waveform to pass through. Capacitor C2 fills in and smoothes out the gaps between the cycles of the RF signal. Only the audio signal passes on to the headphones. Amplification can improve the sensitivity of this receiver, but other receiver types have much better selectivity.

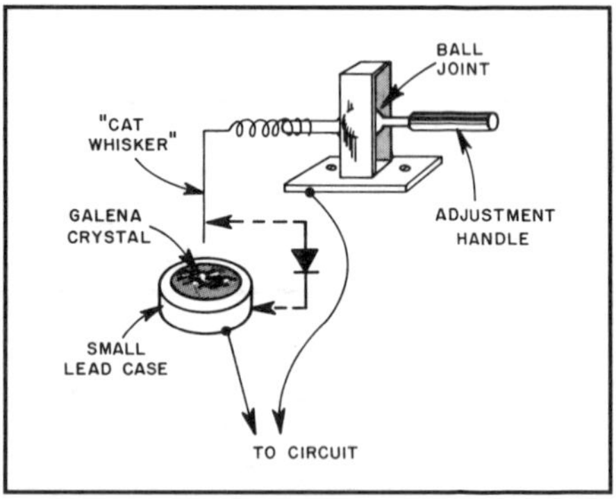

Figure 7-10—Here is an example of a galena crystal and a cat's whisker, the detector in a crystal set.

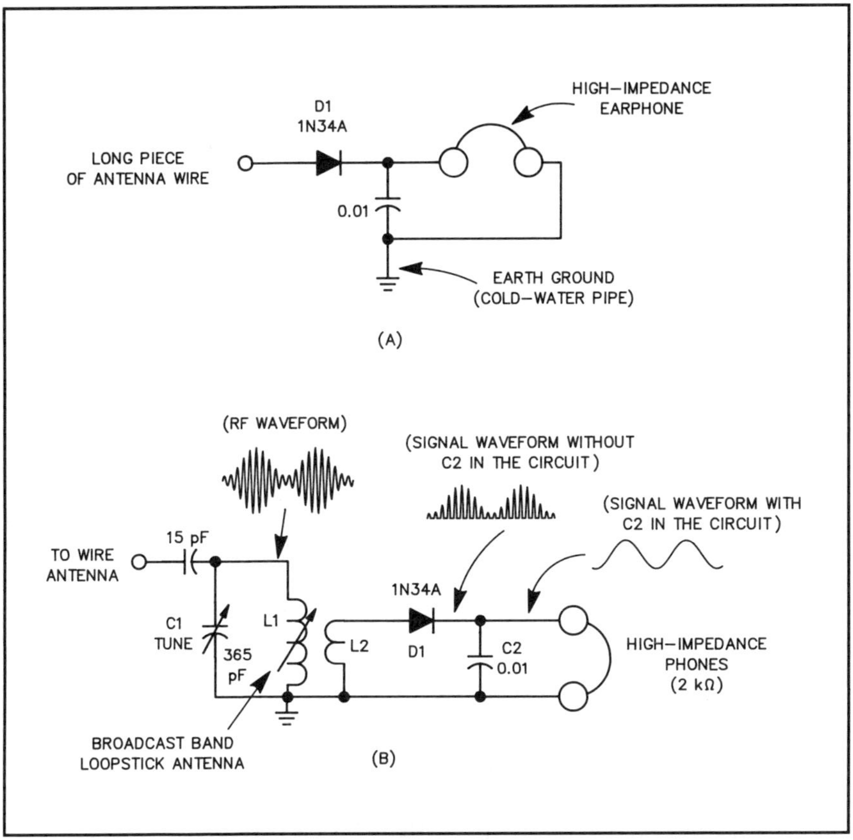

Figure 7-11—Illustration A shows a simple AM receiver using a crystal detector. It consists only of a wire antenna, detector diode, capacitor, earphone and earth ground. The circuit at B has a tuned circuit that helps separate the broadcast-band signals, but otherwise operates the same as the circuit at A.

Direct-Conversion Receivers

The next step up in receiver complexity is shown in Figure 7-12. Direct-conversion means the RF signal is converted directly to audio, in one step. The incoming signal is combined with one from a **variable-frequency oscillator (VFO)** in the *mixer* stage. We mix an incoming signal at 7040 kHz with a signal from the VFO at 7041 kHz. The output of the mixer contains signals at 7040 kHz, 7041 kHz, 14,081 kHz and 1 kHz (the original signals and their sum and difference frequencies). One of these signals (1 kHz) is within the range of human hearing. We use an audio amplifier and hear it in the headphones.

By changing the VFO frequency, other signals in the range of the receiver can be converted to audio signals. Direct-conversion receivers are capable of providing good usable reception with relatively simple, inexpensive circuits. Their main disadvantages are microphonics and ac hum.

Microphonics are caused by vibrating circuit components when the receiver is

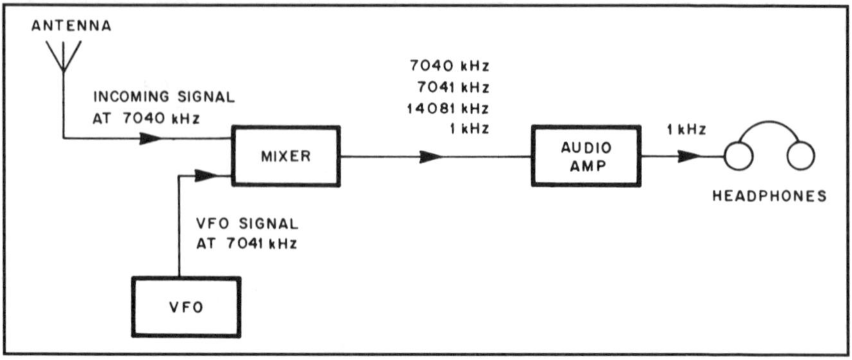

Figure 7-12—A block diagram of a direct-conversion receiver. This receiver converts RF signals directly to audio, using only one mixer.

tuned or adjusted. A receiver must amplify microvolt signals to several volts, so they can be heard on a speaker or in headphones. Direct-conversion receivers do most of their amplification at AF (audio frequencies). A large amount of audio amplification is needed to build up the mixer (detector) output signal. Unfortunately, the movement of components generates signals (like a microphone) that are also amplified. Better receiver designs avoid microphonics by sharing the amplification among RF, IF (intermediate frequency) and AF stages.

Hum is induced from the 60-Hz ac power lines in buildings and houses. The 60-Hz ac can sometimes modulate the VFO signal. You will then hear a hum or buzz in the headphones.

Superheterodyne Receivers

A block diagram of a simple superheterodyne receiver for CW and SSB is shown in Figure 7-13. The first mixer produces a signal at the **intermediate frequency** or **IF**, typically 9 MHz in a modern receiver. All the amplifiers after the first mixer are designed for peak efficiency at the IF. The superhet receiver solves the selectivity/bandwidth problem by converting all signals to the same IF before filtering and amplification. To receive SSB and CW signals, a second mixer, called a product detector, is used. The product detector mixes the IF signal with a signal from the **beat-frequency oscillator (BFO)**. The BFO converts the received-signal information from the intermediate frequency to the audio range. The product-detector output contains audio that can be amplified and sent to a speaker or headphones.

FM Receivers

An FM superheterodyne receiver is similar, but with a few different stages. It has a wider bandwidth filter and a different type of detector. One common FM detector is the **frequency discriminator**. The discriminator output varies in amplitude as the frequency of the incoming signal changes. You can see that it performs the opposite job of the frequency or phase modulator we studied earlier in this chapter.

Most FM receivers also have a **limiter** stage between the IF amplifier and the detector. The limiter makes the receiver less sensitive to amplitude variations and pulse noise than AM or SSB/CW receivers. As its name suggests, the limiter output

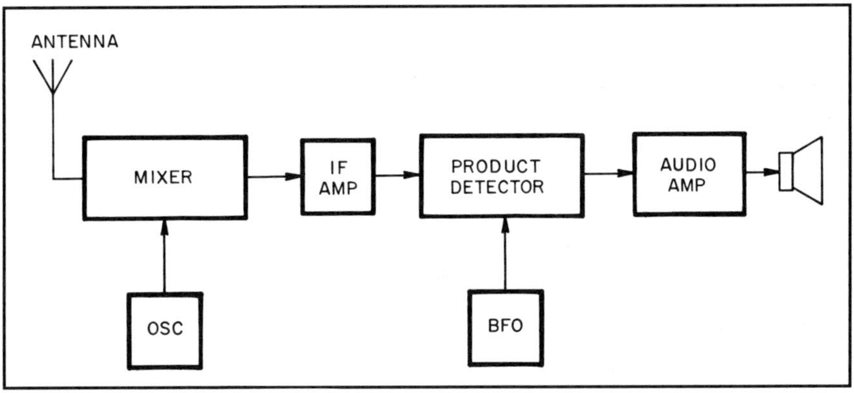

Figure 7-13—A block diagram of a superheterodyne SSB/CW receiver. One mixer converts the incoming signal to the intermediate frequency (IF) and another mixer, called a product detector, recovers the audio or Morse code and converts it to an audio signal.

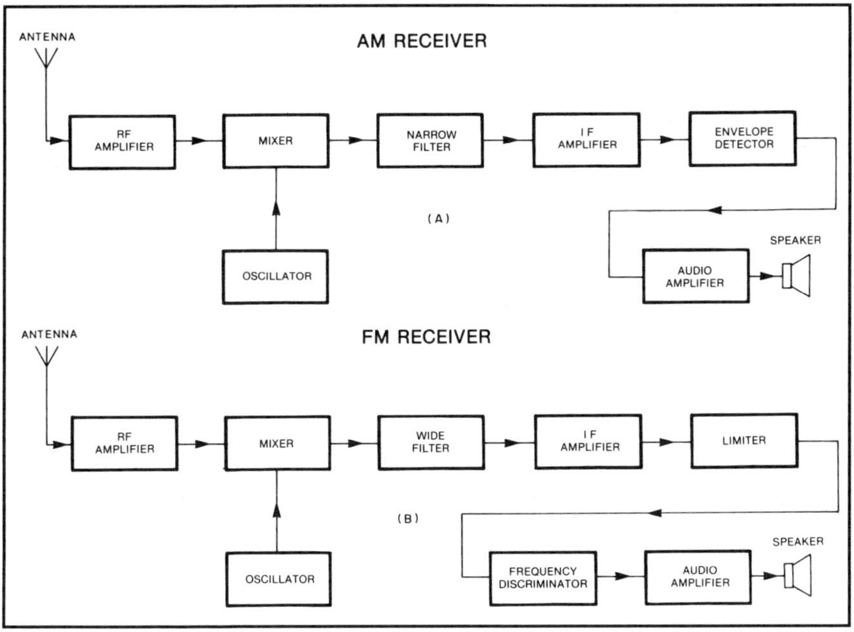

Figure 7-14—AM and FM receivers use similar stages, with a few variations.

remains almost constant when the signal level fluctuates. Noise pulses are also amplitude-modulated signals. The limiter does not pass them on to the detector. This feature makes FM popular for mobile and portable communications. A comparison between an AM receiver and an FM receiver is shown in Figure 7-14.

[This completes your study of this chapter. Now turn to Chapter 10 and study questions T7A04, T7A06, T7A07, T7A09 and T7A10. Review any material you're unsure of before you go on to the next chapter.]

KEY WORDS

Bandwidth—The amount of space (measured in hertz) that a signal occupies.

CW—A communications mode transmitted by on/off keying of a radio-frequency signal. Another name for international Morse code.

Data—Computer-based modes, such as **RTTY** and **packet radio**.

Emission types—Term for the different modes authorized for use on the Amateur Radio bands. Examples are **CW**, **SSB** and **FM**.

Frequency deviation—The instantaneous change in frequency for a given FM or PM signal. Too much deviation (overdeviation or overmodulation) causes **splatter**.

Frequency-modulated (FM) voice—Mode of **voice (phone) communications** used on repeaters.

Frequency-shift keyed telegraphy (FSK)—A method of transmitting radioteletype information by switching an RF carrier between two separate frequencies. FSK RTTY is most often used on HF.

Intermediate-frequency (IF) filter—A circuit that limits the bandwidth of signals fed through the intermediate-frequency (IF) stage of a receiver.

Packet radio—A system of digital communication whereby information is broken into short bursts. The bursts ("packets") also contain addressing and error-detection information.

Phone—Another name for **voice communications**.

Radioteletype (RTTY)—Radio signals sent from one teleprinter machine to another machine. Anything that one operator types on his teleprinter will be printed on the other machine. Also known as *narrow-band direct-printing telegraphy*.

Repeater—An amateur station that receives a signal and retransmits it for greater range.

Sidebands—The sum or difference frequencies generated when an RF carrier is mixed with an audio signal. **Single-sideband** phone (SSB) signals have an upper sideband (USB—that part of the signal above the carrier) and a lower sideband (LSB—the part of the signal below the carrier). SSB transceivers allow operation on either USB or LSB.

Single sideband (SSB)—Emission mode that describes the type of **voice** emission used most on the HF bands.

Splatter—A type of interference to nearby stations. Occurs when a transmitter is overmodulated.

Terminal node controller (TNC)—A device that accepts information from a computer and converts the information into packets. The TNC also receives packets and extracts information to be displayed by a computer.

Test emission—Transmissions containing no information, such as an unmodulated carrier.

Tone-modulated CW (MCW)—A signal produced by feeding an audio tone into the microphone jack of an FM transmitter, and then keying the tone on and off to send Morse code.

Voice communications—Hams can use several voice modes, including **FM** and **SSB**.

8 Signals and Emissions

EMISSION MODES

Hams are authorized to use several different communications modes. Each has advantages for certain types of operating. We'll take a brief look at a few of the most common. Table 8-1 lists the official FCC designators for the emissions authorized for use on the amateur bands.

Table 8-1
Partial List of FCC Emission Designators

1)	*First Symbol—Modulation Type*	
	Unmodulated carrier	N
	Double-sideband full carrier	A
	Double-sideband reduced carrier	R
	Single-sideband suppressed carrier	J
	Vestigial-sideband	C
	Frequency modulation	F
	Phase modulation	G
	Various forms of pulse modulation	P,K,L,M,Q,V,W,X
2)	*Second Symbol—Nature of Modulating Signals*	
	No modulating signal	0
	A single channel containing quantized or digital information without the use of a modulating subcarrier	1
	A single channel containing quantized or digital information with the use of a modulating subcarrier	2
	A single channel containing analog information	3
	Two or more channels containing quantized or digital information	7
	Two or more channels containing analog information	8
3)	*Third Symbol—Type of Transmitted Information*	
	No information transmitted	N
	Telegraphy—for aural reception	A
	Telegraphy—for automatic reception	B
	Facsimile	C
	Data transmission, telemetry, telecommand	D
	Telephony	E
	Television	F

MODULATION

Modulation is the process of varying some characteristic (amplitude, frequency or phase) of an RF carrier wave. The carrier varies in accordance with the instantaneous variations of some external signal to convey information. The radio transmission may be modulated by speech or audio tones. The on-off keying of a CW transmitter with Morse code is a form of modulation. Any method of putting information on an RF carrier is a form of modulation. An unmodulated carrier wave (a steady signal with no information included) is called a **test emission**.

CW (Morse Code)

CW, also called *international Morse code*, is transmitted by on/off switching of a radio-frequency signal. This is demonstrated whenever you use a key to send CW. To send code, you press a lever. To stop sending, you let up on the lever. Code is either on or off. What could be simpler?

It is interesting to note that you can operate Morse code with an FM transceiver. You can feed an audio tone into the microphone jack, and then key the audio tone on and off to send Morse code. This type of signal is **tone-modulated CW (MCW)**.

Single-Sideband Phone (SSB)

Single sideband (SSB) is the most common **voice communications** mode on HF. You can use either the *lower* **sideband** (LSB) or the *upper* **sideband** (USB) to transmit an SSB signal. The carrier and one of the two sidebands are suppressed in an SSB signal. The upper sideband is the part of the signal above the carrier frequency, while the lower sideband is the part of the signal below the carrier. Amateurs normally use upper sideband while operating SSB on 10 meters and on VHF and UHF.

Frequency Modulation (FM) and Phase Modulation (PM)

More hams use **frequency-modulated (FM) voice (phone)** than any other

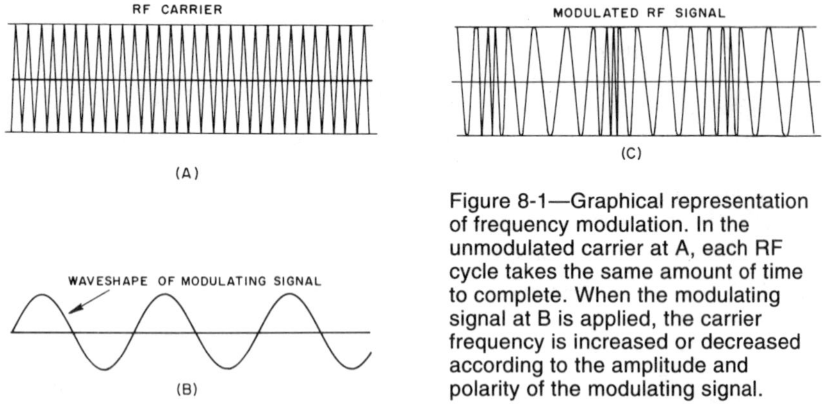

Figure 8-1—Graphical representation of frequency modulation. In the unmodulated carrier at A, each RF cycle takes the same amount of time to complete. When the modulating signal at B is applied, the carrier frequency is increased or decreased according to the amplitude and polarity of the modulating signal.

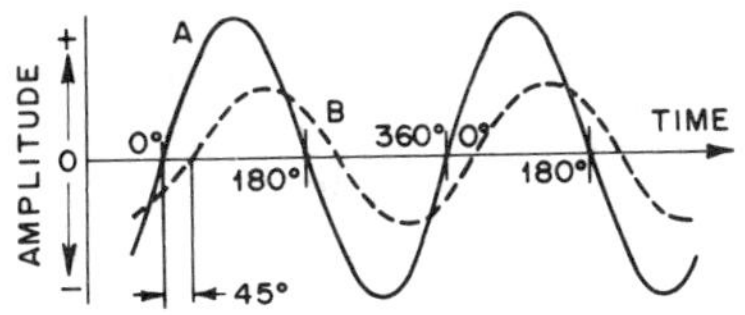

Figure 8-2—When two waves of the same frequency start their cycles at slightly different times, the time difference (or phase difference) is measured in degrees. In this diagram wave B starts 45 degrees (one-eighth cycle) later than wave A, and so lags wave A by 45 degrees. (We can also say that wave A leads wave B by 45 degrees.)

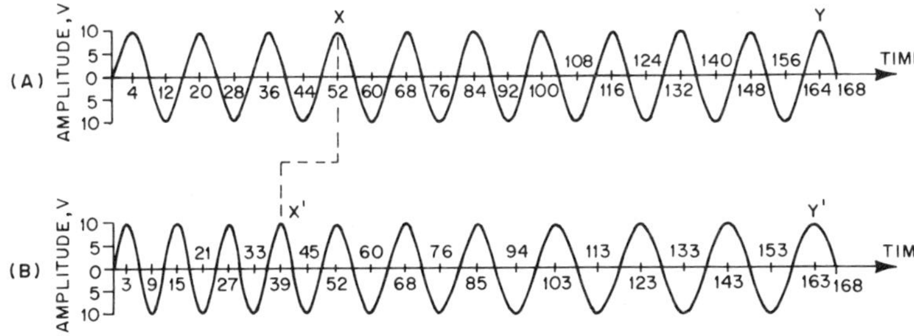

Figure 8-3—Graphical representation of phase modulation. The unmodulated wave is shown at A. After modulation, cycle X' leads cycle X, and all the cycles to the left of cycle X' are compressed. To the right, the cycles are spread out.

communications mode. FM is produced when the frequency of the carrier current is varied in accordance with variations in a modulating signal. See Figure 8-1. VHF and UHF FM phone (voice) operation takes two forms: simplex and repeater. *Simplex* operation means the stations are talking to each other directly, on the same frequency. This is similar to making a contact on the HF bands. **Repeaters** amplify and relay incoming signals and transmit them on a different frequency.

FM is used on VHF/UHF simplex and repeaters because of its high-fidelity audio and high signal-to-noise ratio. This makes FM especially suited to channelized local communication. The signals must be above a minimum level, however. SSB and CW are much better suited to weak-signal operation.

The main disadvantage of FM is its wide **bandwidth**. *Phase modulation (PM)* is produced when the phase of the carrier current changes in accordance with va- tions in a modulating signal. See Figures 8-2 and 8-3. Phase modulation ar quency modulation produce signals that are very similar.

Radioteletype (RTTY)

Radioteletype (RTTY) communication using the Baudot code (also known as the International Telegraph Alphabet number 2, or ITA2) is used on the Amateur Radio HF bands in most areas of the world. The Baudot code represents each character with a string of five bits of digital information. Each character has a different combination of bits. There are only 32 possible Baudot code combinations. This limits the number of possible characters. All text is in upper-case characters. To provide numbers and punctuations, you shift between the letters case (LTRS) and the figures case (FIGS).

RTTY communication is similar to CW, in that it uses two states to convey information. Instead of switching a carrier on and off, the carrier is left on continuously and switched between two different frequencies. The transmitter carrier frequency "shifts" between two frequencies, called MARK and SPACE. MARK is the ON state; SPACE is the OFF state. This is called **frequency-shift keyed telegraphy (FSK)**. On 10-meter RTTY, the MARK and SPACE frequencies are normally 170 Hz apart. This is known as 170-Hz shift RTTY. On VHF, *audio-frequency shift keying (AFSK)* is used. With AFSK, two switched audio tones are fed into the microphone input of an FM transmitter.

Packet Radio

Packet radio is a type of computer-based **data** communications whereby information is broken into short bursts. In addition to the information from your computer, each packet contains addressing, error-checking and control information. The addressing information includes the call sign of the station that sent the packet, and the call sign of the station the packet is being sent to. The address may also include call signs of stations that are being used to relay the packet. The error-checking information allows the receiving station to determine whether the received packet contains any errors. If the received packet contains errors, the receiving station waits until the transmitting station sends it again. Breaking up the data into small parts allows several users to share the channel. Packets from one user are transmitted in the spaces between packets from other users.

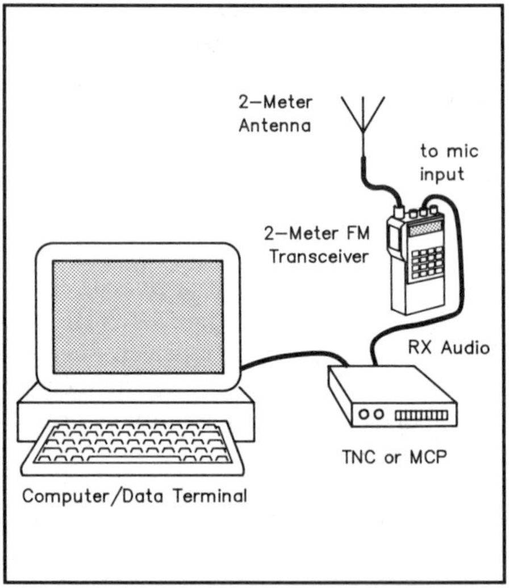

Figure 8-4—A terminal node controller connects between your transceiver and your computer. Multimode communications processors (MCPs) can operate packet and several other modes.

All you need to set up a VHF packet-radio station is a VHF or UHF FM transceiver (with an antenna), a computer or ASCII terminal and a **terminal node controller (TNC)** or multimode controller (which allows operation on several modes). The TNC connects between the transceiver microphone input and the computer. See Figure 8-4.

[Now turn to Chapter 10 and study questions T8A01 through T8A05, T8A07, T8A08, T8A11, T8B04 and T8B05. Review this section if you have difficulty.]

BANDWIDTH

The amount of space in the radio-frequency spectrum that a signal occupies is called its **bandwidth**. The bandwidth of a transmission is determined by the information rate. Thus, a pure, continuous, unmodulated carrier has a very small bandwidth with no sidebands. A television transmission, which contains a great deal of information, is several megahertz wide.

Receiver Bandwidth

Receiver bandwidth determines how well you can receive one signal in the presence of another signal that is very close in frequency. The enjoyment you'll experience will depend greatly on how well you can isolate the signal you are receiving from all the others nearby.

Bandwidth is a measure of *selectivity*: how wide a range of frequencies is received with the receiver tuned to one frequency. For example, if you can hear signals as much as 3 kHz above and 3 kHz below the frequency to which you are tuned, your receiver has a bandwidth of at least 6 kHz. If you cannot hear signals more than 200 Hz above or below the frequency to which you are tuned, the bandwidth is only 400 Hz. The narrower the bandwidth, the greater the selectivity and the easier it is to copy one signal when there is another one close by in frequency.

Selectivity is determined by special **intermediate-frequency (IF) filters** built into the receiver. Some receivers have several filters so you can choose different bandwidths. They are necessary because different **emission types** occupy a wider frequency range than others. A 250-Hz-bandwidth filter is excellent for separating CW signals on a crowded band, but it's useless for listening to SSB or FM transmissions. For those modes, a wider filter is needed to allow all the transmitted information to reach the detector.

Most receivers designed for single-sideband voice operation come standard with a filter selectivity of around 2.8 kHz. This is ideal for SSB. Although this level of selectivity is also usable on CW, a narrower bandwidth is needed to prevent adjacent CW signals from getting through at the same time.

As we have seen in these examples, radioteletype emissions are wider than CW signals, and SSB signals are even wider than that. Next, we will consider FM and PM, which can occupy even more bandwidth.

Bandwidth In FM and PM

Frequency deviation is the instantaneous change in frequency for a given signal. The frequency swings just as far in both directions, so the total frequency swing is equal to twice the deviation. In addition, there are sidebands that increase the bandwidth still further. A good estimate of the bandwidth is twice the maximum frequency deviation plus the maximum modulating audio frequency:

$Bw = 2 \times (D + M)$ (Equa

where:

Bw = bandwidth
D = maximum frequency deviation
M = maximum modulating audio frequency

An FM transmitter using 5-kHz deviation and a maximum audio frequency of 3 kHz uses a total bandwidth of about 16 kHz. The actual bandwidth of a typical FM signal is somewhat greater than this, but it is a good approximation.

Frequency Deviation

With direct FM, frequency deviation is proportional to the amplitude of the modulating signal. With phase modulation, the frequency deviation is proportional to both the amplitude and the frequency of the modulating signal. Frequency deviation is greater for higher audio frequencies. This is actually a benefit, rather than a drawback, of phase modulation. In radio communications, speech frequencies between 300 and 3000 Hz need to be reproduced for good intelligibility. In the human voice, however, speech sounds between 2000 and 3000 Hz are not as strong as at lower frequencies. Something must be done to increase the amplitude of these frequencies when a direct FM transmitter is used. A circuit called a preemphasis network amplifies the sounds between 2000 and 3000 Hz. With phase modulation, the preemphasis network is not required, because the deviation already increases with increasing audio input frequency.

Modulation Index

The sidebands that occur from FM or PM differ from those resulting from AM. With AM, only a single set of sidebands is produced for each modulating frequency. FM and PM sidebands occur at integral multiples of the modulating frequency on either side of the carrier, as shown in Figure 8-5. Because of these multiple sidebands,

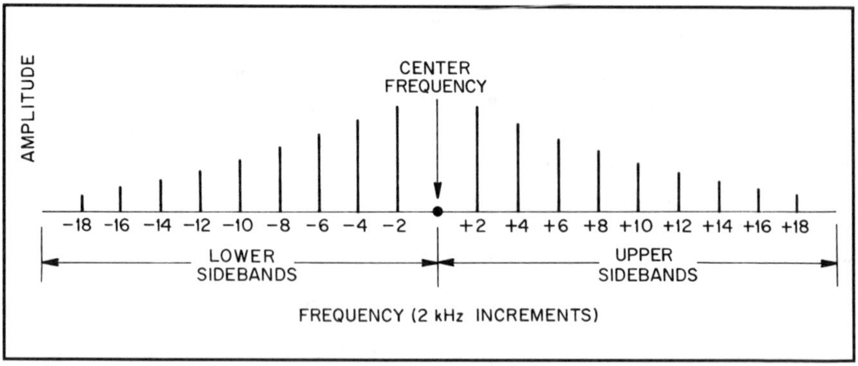

Figure 8-5—When an RF carrier is frequency-modulated by an audio signal, multiple sidebands are produced. FM sidebands are spaced above and below the center frequency of the carrier. This drawing shows a group of sidebands resulting from the application of a 2-kHz tone to the transmitter modulator. The sidebands farthest from the center frequency are the lowest in amplitude. The sidebands nearest to the center frequency can be considered the significant sidebands, their amplitudes being fairly large.

FM or PM signals inherently occupy a greater bandwidth. The additional sidebands depend on the relationship between the modulating frequency and the frequency deviation. The ratio between the peak carrier frequency deviation and the audio modulating frequency is called the *modulation index.*

Given a constant input level to the modulator, in phase modulation, the modulation index is constant regardless of the modulating frequency. For an FM signal, the modulation index varies with the modulating frequency. In an FM system, the ratio of the maximum carrier-frequency deviation to the highest modulating frequency is called the *deviation ratio*. The modulation index is a variable that depends on a set of operating conditions, but the deviation ratio is a constant. The deviation ratio for narrow-band FM is 5000 Hz (maximum deviation) divided by 3000 Hz (maximum modulation frequency) or 1.67.

Frequency multiplication offers a means for obtaining any desired amount of frequency deviation, whether or not the modulator is capable of that much deviation. Overdeviation (overmodulation) of an FM transmitter causes **splatter**, out-of-channel emissions that can cause interference to adjacent frequencies.

A frequency-multiplier stage is an amplifier that produces harmonics of the input signal. By using a filter to select the desired harmonic, the proper output frequency is obtained. A frequency multiplier is not the same as a mixer stage, which is used to obtain the desired output frequency in an SSB or CW rig. A mixer shifts, or translates, an oscillator frequency to some new frequency.

[Now turn to Chapter 10 and study questions T8B06 through T8B11. Review this section if you have difficulty.]

KEY WORDS

Antenna—A device that picks up or sends out radio waves.

Balanced antenna—An antenna that is symmetrical about the feed point. There are two equal sides to the antenna. A dipole is a balanced antenna.

Balun—Contraction for *bal*anced to *un*balanced. A device to couple a balanced load to an unbalanced source, or vice versa.

Beam antenna—A directional antenna. A beam antenna must be rotated to provide coverage in different directions.

Coaxial cable—Coax (pronounced kó-aks). A type of feed line with one conductor inside the other.

Cubical quad antenna—An antenna built with its elements in the shape of four-sided loops.

Delta loop antenna—A variation of the cubical quad with triangular elements.

Director—An element in front of the driven element in a Yagi and some other directional antennas.

Driven element—The part of an antenna that connects directly to the feed line.

Feed line (feeder)—The wires or cable used to connect a transmitter or receiver to an antenna. (Also called *transmission line.*)

Forward power—The power of the radio signal moving from the transmitter to the antenna along a **feed line**. (Also see **reflected power**.)

Groundplane antenna—A vertical antenna mounted above ground, and with horizontal radial wires or a metal surface to form a simulated ground under the base of the antenna.

Horizontally polarized wave—An electromagnetic wave with its electric lines of force parallel to the ground.

Nondirectional antenna—An antenna that sends radio signals equally well in all compass directions. A single vertical antenna, such as a **groundplane antenna** is an example of a nondirectional or *omnidirectional* antenna.

Open-wire feed line—Parallel-conductor feeder with air as its primary insulation material.

Parallel-conductor feed line—Feed line with two conductors held a constant distance apart.

Parasitic element—Part of a directive antenna that derives energy from mutual coupling with the driven element. Parasitic elements are not directly connected to the feed line.

Polarization—Describes the electrical-field characteristic of a radio wave. An antenna that is parallel to the surface of the earth, such as a dipole, produces horizontally polarized waves. One that is perpendicular to the earth's surface, such as a quarter-wave vertical, produces vertically polarized waves.

Reflected power—The power of the radio signal moving from the antenna to the transmitter along a **feed line**. The power of the signal *reflected* from the antenna because of an impedance mismatch. (Also see **forward power**.)

Reflector—An element behind the driven element in a Yagi and some other directional antennas.

RF burn—A flesh burn caused by exposure to a strong field of radio-frequency (RF) energy.

Standing-wave ratio (SWR)—Sometimes called VSWR. A measure of the impedance match between the feed line and the antenna. Also, with a Transmatch in use, a measure of the match between the feed line from the transmitter and the antenna system. The system includes the Transmatch and the line to the antenna. VSWR is the ratio of maximum voltage to minimum voltage along the feed line. Also the ratio of antenna impedance to feed-line impedance when the antenna is a purely resistive load.

Unbalanced line—Feed line with one conductor at ground potential, such as coaxial cable.

Vertically polarized wave—A radio wave that has its electric lines of force perpendicular to the surface of the earth.

Yagi antenna—The most popular type of amateur directional (beam) antenna. It has one driven element and one or more additional elements.

9 Antennas and Feed Lines

We know that a transmitter generates radio-frequency energy. We convert this electrical energy into radio waves with an antenna. An antenna converts current into an electromagnetic field (radio waves). The radio waves spread out or *propagate* from the antenna. You might relate their travel to the ever-expanding waves you get when you drop a pebble in water. Waves from an antenna radiate in all directions, though, not just in a flat plane.

An antenna system radiates a signal from your transmitter. Antenna systems include a number of items. The antenna itself (the part that actually radiates the signal), and the **feed line** (sometimes called *transmission line*) are the most important parts. Any coupling devices or matching networks are also included. These devices transfer power from the transmitter to the line and from the line to the antenna. In some simple systems the feed line may be part of the antenna. Coupling devices might not be used there. For your Technician exam, you should know the basics of antennas and feed lines commonly used in Amateur Radio.

FEED-LINE ATTENUATION

We use a **feed line** (or *transmission line*) to transfer power from a transmitter to an antenna. There are two common types of feed lines: **parallel-conductor feed line** and **coaxial cable**. Both can be constructed in a variety of forms. We can divide each of these feed-line types into two classes: those in which the majority of the space between the conductors is air, and those in which the conductors are embedded in and separated by a solid plastic or foam insulation (*dielectric*).

Air-Insulated Feed Lines

Air-insulated feed lines have the lowest loss per unit length (usually expressed in dB/100 ft). Adding a solid dielectric between the conductors increases the losses in the feed line. The power loss causes heating of the dielectric. (*Dielectric* is the technical term for an insulating material.) As frequency increases, conductor and dielectric losses become greater for coaxial and parallel-conductor feed lines. An advantage of this type of transmission line is that it can be operated at a high **standing-wave ratio (SWR)** and still retain its low-loss properties. (SWR is discussed in more detail later in this chapter.)

Figure 9-1 shows typical construction techniques for air-insulated parallel-conductor feed lines. Insulating rods or spacers hold the wires a fixed distance apart. These spacers may be made from phenolic, polystyrene or Lucite. Their length may be between 1 and 6 inches to hold the conductors a small fraction of a wavelength apart. Shorter spacers are more common for the higher frequencies. These spacers are placed some distance along the line to keep the wires parallel to each other. This type

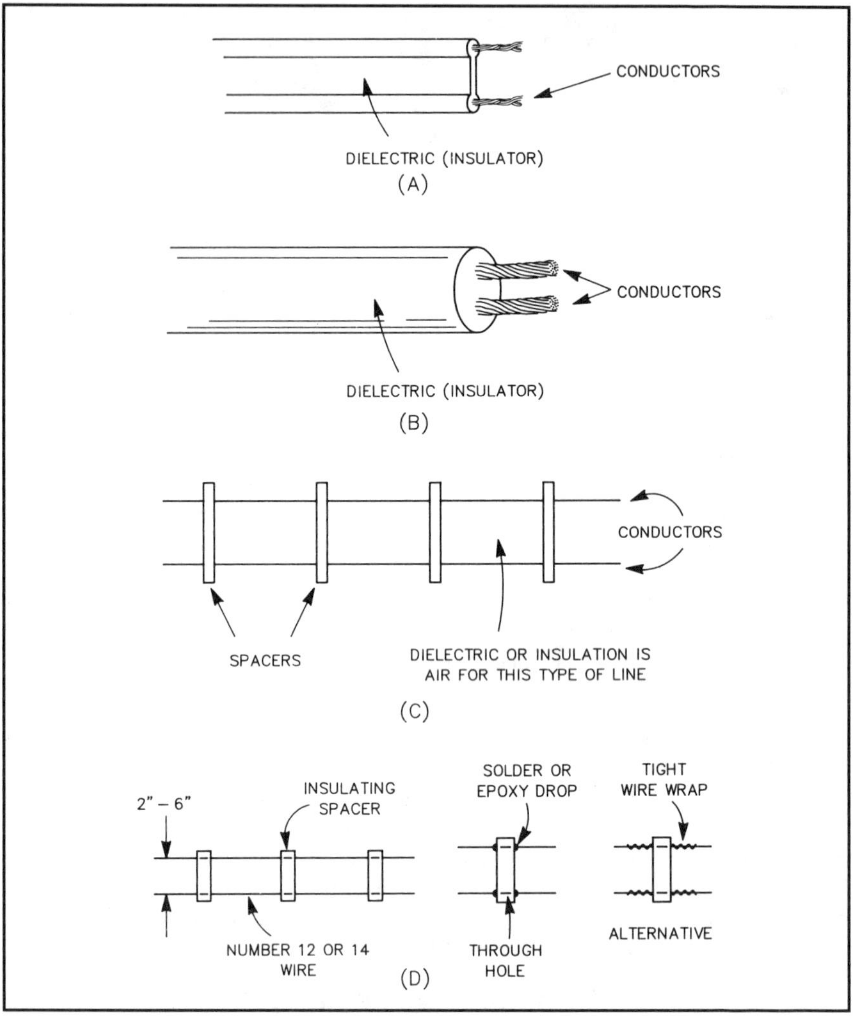

Figure 9-1—Parallel-conductor feed lines. At A, common 300-ohm TV twin lead. B shows 72-ohm transmitting twin lead, designed to carry higher power levels. C shows parallel line that uses air as the dielectric, with insulating spacers every few inches. Ladder line is its common name. D shows the construction of home-built ladder line.

of line is often called **open-wire feed line**. Open-wire line has the least loss of any feed line type commonly used by amateurs. This means for the same feed-line length, more of your transmitter power will get to your antenna than for any other feed-line type.

Characteristic impedance (Z_0) is the opposition to electric current that an antenna feed line presents. Impedance includes factors other than resistance, and applies to alternating currents. Ideally, the characteristic impedance of a feed line is the same as the transmitter output impedance and the antenna input impedance.

The characteristic impedance of an air-insulated parallel-conductor line depends on the diameter of the wires used in the feed line and the spacing between them. The greater the spacing between the conductors, the higher the characteristic impedance of the feed line. The impedance decreases, however, as the size of the conductors increases. The characteristic impedance of a feed line is not affected by the length of the line.

Solid-Dielectric Feed Lines

Transmission lines in which the conductors are separated by a flexible dielectric have several advantages over air-dielectric line: They are less bulky, maintain more uniform spacing between conductors, are generally easier to install and are neater in appearance. Both parallel-conductor and coaxial lines are available with this type of insulation.

One disadvantage of these types of lines is that the power loss per unit length is greater than air-insulated lines because of the dielectric. As the frequency increases, the dielectric losses become greater. The power loss causes heating of the dielectric. Under conditions of high power or high SWR, the dielectric may actually melt and cause short circuits or arcing inside the line.

TV-type parallel-conductor line, commonly called twin lead, is perhaps the best-known form of solid-dielectric parallel-conductor feed line. Twin lead consists of two number 20 wires that are molded into the edges of a polyethylene ribbon about a half-inch wide. The presence of the solid dielectric lowers the characteristic impedance of the line as compared to the same conductors in air.

The fact that part of the field between the conductors exists outside the solid dielectric leads to operating disadvantages. Dirt or moisture on the surface of the ribbon tends to change the characteristic impedance. Weather effects can be minimized, however, by coating the feed line with silicone grease or car wax. In any case, the changes in the impedance will not be very serious if the line is terminated in its characteristic impedance (Z_0). If there is a considerable standing-wave ratio, however, then small changes in Z_0 may cause wide fluctuations of the input impedance.

Parallel-conductor feed lines are *balanced lines* because neither conductor connects to ground. They can be connected directly to **balanced antennas**, which are symmetrical about the feed point.

Coaxial Cable

Coaxial cable has one conductor inside the other. It's like a wire inside a flexible tube. The center conductor is surrounded by insulation, and the insulation is surrounded by a wire braid called the shield. The whole cable is then encased in a tough vinyl outer coating, which makes the cable weatherproof. Coax comes in a variety of sizes, with different electrical properties. Figure 9-2 shows the construction of several types of coaxial cables used by amateurs.

The most common types of coax have either a 50-ohm or 72-ohm *characteristic impedance*. Coax designated RG-58, RG-8 and RG-213 are 50-ohm cables. Some coax designations may also include a suffix such as /U, A/U or B/U, or bear the label "polyfoam." Feed line of this type may be used with most antennas. Cables labeled RG-59 or RG-11 are 72-ohm lines. Many hams use these types to feed dipole antennas.

Any line that feeds an antenna absorbs a small amount of transmitter power. That radio-frequency energy is lost, because it serves no useful purpose. (The energy is converted to heat, and warms the feed line slightly.) The loss occurs because the

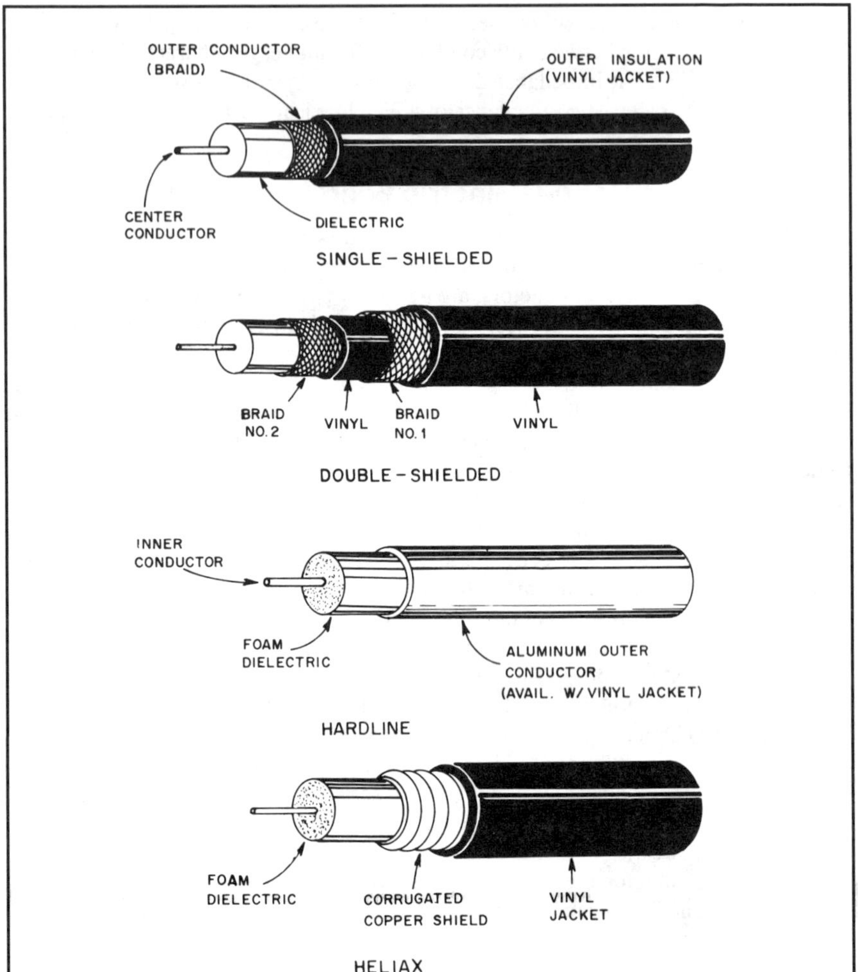

Figure 9-2—Types of coaxial cable used by amateurs. Abbreviated coax, it has a center conductor surrounded by insulation. The second conductor, called the shield, goes around that. Plastic insulation covers the entire cable.

wires are not perfect conductors, and the insulating material is not a perfect insulator. Signal loss also increases slightly when the SWR is greater than 1:1, so we try to keep the SWR below 2:1 if possible. If you notice that your feed line is hot after you transmit for a few minutes, it is an indication your feed-line loss may be high. It may also mean the SWR on the line may be too high.

Better-quality coaxial cables have lower loss than poor-quality cables. More of the transmitter power (radio energy) is lost as heat in a poor-quality coaxial cable. You usually get better quality coax if you stick with name brands. You can also look at the shield braid on the coax. Better-quality cables have more complete coverage of the center insulator. If you can easily see through the holes in the braid, you should probably select another cable.

The larger coax types, RG-8, RG-213 and RG-11, have less signal loss than the smaller types. If your feed line is less than 100 feet long, you probably won't notice the small additional signal loss, at least on the HF bands. This, combined with light weight and flexibility, is why many HF operators find the smaller coax better suited to their needs. Also, the smaller feed line costs about half as much per foot as the larger types.

On the VHF/UHF bands, however, you will find the losses in RG-58 and RG-59 more noticeable, especially if your feed line is longer than about 50 feet. On these bands, most amateurs use RG-213 coax or even lower-loss coaxial cables like Hardline or Heliax.

The larger the diameter, the higher the power capability of the line because of the increased dielectric thickness and conductor size. In general, losses decrease as the cable diameter increases, because there is less power lost in the conductor.

Amateurs commonly use RG-8, RG-58, RG-174 and RG-213 coaxial cable. RG-8 and RG-213 are similar cables, and they have the least loss of the types listed here. RG-174 is only about ⅛ inch in diameter, and it has the highest loss of the cables listed here.

As the frequency increases, the conductor and dielectric losses become greater, causing more attenuation of the signal in the cable. Although the cost, size and weight are larger for RG-213, it is usually the best choice (of the cables mentioned) for a run of over 150 feet for frequencies up to 54 MHz (the amateur 6-meter band).

Of the common coaxial cables discussed here, RG-213 has the least loss. RG-58 has a bit more loss than RG-213, and RG-174 has the most loss of these cables. RG-174 is normally used for cables that connect sections of a transmitter or receiver, or for short interconnecting cables in a low-power system. Some amateurs use RG-174 cable as the feed line for a low-power portable station.

Extra cable length increases attenuation. When using coaxial cable, you should try to use a matched antenna and feed line. You should then be able to change feed line lengths without significantly affecting the antenna system. Your feed line has only to be long enough to reach your antenna. A low SWR on the line means that the impedance "seen" by the transmitter will be about the same regardless of line length. You can cut off or shorten excess cable length to reduce attenuation of the signal caused by feed-line loss. (This does not apply to multiple antennas in phased arrays or line sections used for impedance-matching purposes.)

Attenuation is not affected by the characteristic impedance of a matched line if the spacing between the conductors in the coaxial cable is a small fraction of a wavelength at the operating frequency.

Coaxial-cable connectors are an important part of a coaxial feed line. Your choice of connectors normally depends on the matching connectors on your radios. Most HF radios and many VHF radios use SO-239 connectors. The mating connector is called a PL-259. (These are military-type designations.) The PL-259 is sometimes called a *UHF connector,* although they are not the best choice for the UHF bands. Figure 9-3 shows the SO-239 connector and its mating PL-259.

Many VHF and UHF hand-held radios use BNC connectors. These connectors are designed for use with RG-58 coax. They produce a low-loss connection that is also weatherproof. Figure 9-4 shows a pair of BNC connectors. BNC connectors are well-suited for use with hand-held-radio antennas because they require only a quarter turn

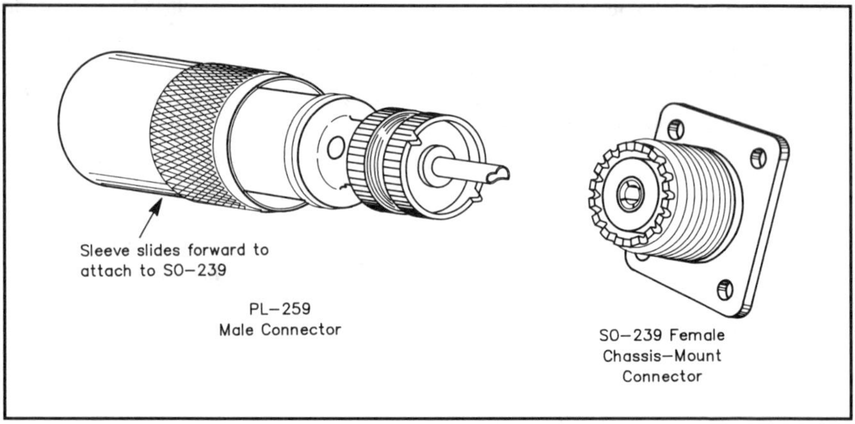

Figure 9-3—This drawing shows a pair of "UHF connectors." They are better known as PL-259 and SO-239 connectors. These connectors are most-often used for HF connections, and are not particularly good at UHF, in spite of their name.

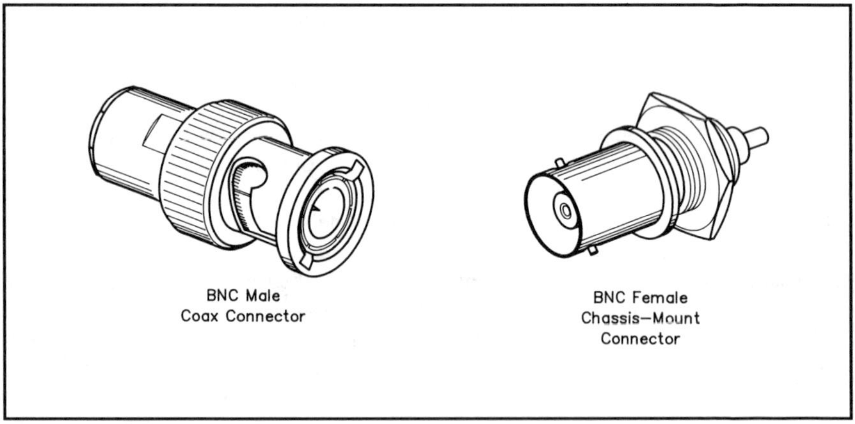

Figure 9-4—BNC connectors are popular VHF connectors. Many hand-held radios use these connectors.

to install or remove, yet they lock securely in place.

Many amateurs prefer N-type connectors for UHF equipment. N connectors are low-loss, high-power connectors. They are designed for use with RG-213 or RG-8 coaxial cable. N connectors have the lowest loss of the common coaxial connectors, which is why they are often used at UHF. Figure 9-5 shows a set of N connectors.

It is a good idea to check your coaxial connectors on a regular basis. Be sure they are clean and tight. If you suspect a bad solder connection, you should resolder the joints. This will help keep the connector resistance at a minimum.

[Turn to Chapter 10 and study questions T9B09, T9C01 through T9C08, and T9C11 before continuing with this chapter. Review this section if needed.]

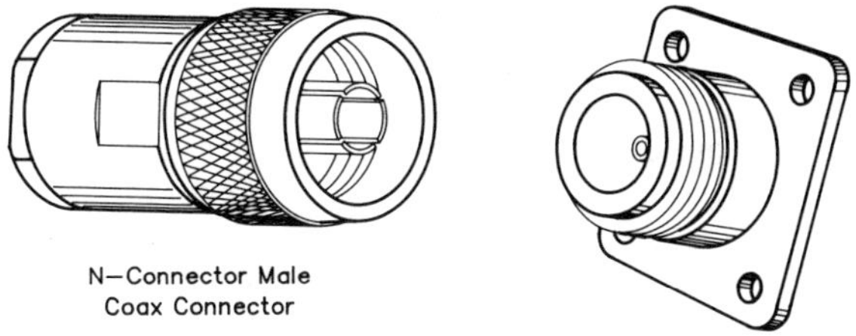

Figure 9-5—N connectors are popular for connecting a coaxial cable to a UHF rig.

BALUNS

A center-fed wire with open ends, such as a dipole antenna, is a **balanced antenna**. In a balanced, center-fed antenna, the current flowing into one half of the antenna is equal to the current in the other half. The two currents are also opposite in *phase*. (*Phase* refers to the relative positions of two points on a wave, or on two different waves at a particular instant of time. If the two currents are opposite in phase, one is in the positive half of the cycle when the other is in the negative half cycle.) You can think of a balanced antenna as one where neither side connects to ground. A balanced antenna is balanced with respect to ground.

If we feed the antenna with a parallel-conductor line, we keep balance throughout the system. Parallel-conductor feed line is balanced: Neither side connects to ground. You can connect this feed line directly to a dipole antenna.

The outside shield of coaxial cable is normally connected to ground. Coax, then, is an **unbalanced line**. When connecting coaxial *(unbalanced)* feed lines to *balanced antennas*, many hams prefer to use a *balun*.

If we feed a dipole at the center with coax, we upset the system balance. One side of the antenna connects to the coax inner, or center, conductor. The other side connects to the coax shield. The shield connects to ground, so coax is an unbalanced line. This unbalanced condition may allow some antenna current to flow down the outside of the coax braid from the antenna. This can lead to several antenna problems, and should be avoided.

One way to feed a dipole antenna with coaxial line is to use a device called a **balun**. Balun is a contraction for *bal*anced to *un*balanced. You install the balun at the antenna feed point, between the coax and the antenna.

Different types of baluns are commercially available, or you can make your own. A common type is a balun transformer, which uses wires wound on a toroidal core. Besides providing a balance, these baluns can transform an impedance, such as from 50 to 75 ohms. Another type is a bead balun. Several ferrite beads go over the outside of the coax, one after the other. The beads tend to choke off any RF current that might otherwise flow on the outside of the shield.

You can easily make another type of choke balun from the coax itself. At the antenna feed point, coil up 10 turns of coax into a roll about 6 inches in diameter. Tape the coax turns together. The inductance of the coiled turns tends to choke off RF currents on the shield.

Feeding a dipole with parallel-conductor line doesn't assure a balanced antenna. You must also consider how the line connects to the transmitter. Most transmitters have a coaxial-style connector. Suppose you are using 450-ohm parallel-conductor feed line. To feed such a balanced antenna system, you would need to install a balun at the *transmitter* end of the line (or use a special balanced matching network). Unless you use a balanced matching network, you need a balun at the transmitter end of the line if you are using any type of parallel-conductor feed line.

[Now turn to Chapter 10 and study questions T9B10 and T9B11. Review this section if you have any difficulty.]

STANDING-WAVE RATIO (SWR)

If an antenna system does not match the characteristic impedance of the transmitter, some of the transmitter energy is reflected from the antenna. The power traveling from the transmitter to the antenna is called **forward power**. When that power reaches the antenna in an unmatched system, some of the power is reflected back down the feed line toward the transmitter. Some of the power is also radiated from the antenna, which is what you want to happen. The power that returns to the transmitter from the antenna is called **reflected power**.

The forward power and the reflected power passing each other on the feed line cause voltage standing waves on the line. When this happens, the RF voltage and current are not uniform along the line. An *SWR meter* measures the voltage **standing-wave ratio (SWR)**. This is the ratio of the maximum voltage on the line to the minimum voltage. (These two points will always be ¼ λ apart.) Lower SWR values mean a better impedance match exists between the transmitter and the antenna system. If a perfect match exists, the SWR is 1:1. Your SWR meter thus gives a relative measure of how well the antenna system impedance matches that of the transmitter.

Modern transmitters are designed to match 52-ohm coaxial lines and antennas. Most commercial antennas are designed to have nearly the same characteristic impedance *when properly adjusted*. So, if your SWR is not close to 1:1, it means your antenna is not properly adjusted for the frequency you are using. We adjust the antenna for minimum SWR (not always 1:1) somewhere in the middle of the band of interest. On other frequencies the SWR may be higher. If the antenna is properly assembled, an SWR of 2:1 or less is probably all right.

If you are using a matching network, you can probably adjust it so the SWR meter reads 1:1. The SWR on the transmission line between the tuner and the antenna will, however, be different. A matching network can cover up the mismatch, but it does not eliminate it.

[Now it is time to turn to Chapter 10 and study questions T9B06, T9B07 and T9B08. Review this section if these questions give you any difficulty.]

VERTICAL ANTENNAS

The **quarter-wavelength vertical antenna** is very popular. It is easy to build at home, requires only one support and can be very effective. On the HF bands (80 to 10 meters) it is often used for DX work. Vertical antennas send radio energy equally well in all compass directions. This is why we sometimes call them **nondirectional antennas** or *omnidirectional antennas*. They also tend to concentrate the signals towards the horizon. Vertical antennas do not generally radiate strong signals straight up, like horizontal dipoles do. Because they concentrate the signals towards the horizon, we say they have *gain* as compared to a dipole. Gain always refers to a comparison with another antenna. A dipole is one common comparison antenna for stating antenna gain.

The $\frac{1}{4}$-λ vertical uses radials under the antenna to reduce ground losses. For operation on 80 through 10 meters, the vertical may be at ground level, and the radials placed on the ground. The key to successful operation with a ground-mounted vertical antenna is a good radial system. The best radial system uses *many* ground radials. Lay them out like the spokes of a wheel, with the vertical at the center. Some hams have buried ground radial systems containing over 100 individual wires.

Ideally, these wires would be $\frac{1}{4}$ λ long or more at the lowest operating frequency. With such a system, earth or ground losses will be negligible.

If you place the vertical above ground, you reduce earth losses drastically. Here, the wires should be cut to $\frac{1}{4}\lambda$ for the band you plan to use. Above ground, you need only a few radials—two to five work well, and four is common. If you install a multiband vertical antenna above ground, use separate ground radials for each band you plan to use. These lengths are more critical than for a ground-mounted vertical. For elevated verticals, you should have a minimum of two radials for each band. You can mount a vertical on a pipe driven into the ground, on the chimney or on a tower.

The radials at the bottom of a vertical antenna mounted above ground form a surface that acts like the ground under the antenna. These antennas are sometimes called **groundplane antennas**. Groundplane antennas are popular because they are easy to build at home and they work well. Figure 9-6 shows construction details of a 2-meter groundplane antenna.

Groundplane antennas are popular mobile antennas because the car body can serve as the ground plane. You can place a magnetic-base whip antenna on the roof of your car, for example. This type of antenna is often used on the VHF and UHF bands for FM voice communication. Because it is a vertical antenna, it is a **nondirectional antenna**, sending radio energy out equally well in all compass directions.

[Turn to Chapter 10 now. Study questions T9A09, T9A10 and T9A11. Review this section if you have trouble answering any of these questions.]

PARASITIC BEAM ANTENNAS

Beam antennas tend to concentrate their transmitted signals in one compass direction. Such antennas must be rotated or turned to provide coverage in different directions. A beam antenna usually has several parts, or elements. In most multiple-element antennas, the additional elements are not directly connected to the feed line. They receive power by mutual coupling from the **driven element**. The driven element is the element connected to the feed line. The additional elements then reradiate the power in the proper phase relationship. Proper phasing achieves *gain* or *directivity* over a simple half-wavelength dipole. These elements are called **parasitic elements**. Beam antennas that use parasitic elements are sometimes called *parasitic beam antennas*.

There are two types of parasitic elements. A **director** is generally shorter than the driven element. A director is located at the front of the antenna. A **reflector** is generally longer than the driven element. A reflector is located at the back of the

Figure 9-6—Simple groundplane antenna for the 146, 222 and 440-MHz bands. The vertical element and radials are 3/32- or 1/16-in. brass welding rod. Although 3/32-in. rod is preferred for the 146-MHz antenna, no. 10 or 12 copper wire can also be used.

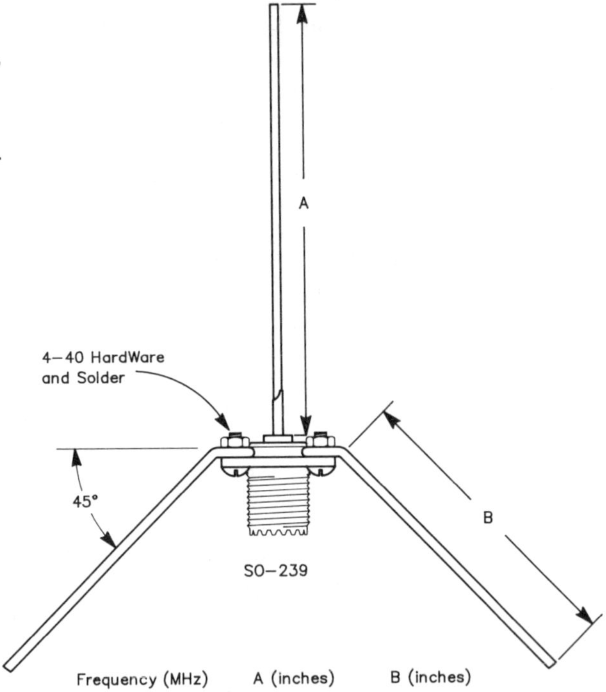

Frequency (MHz)	A (inches)	B (inches)
146	19–5/16	18–11/16
225	12–5/8	12
445	6–38	5–3/4

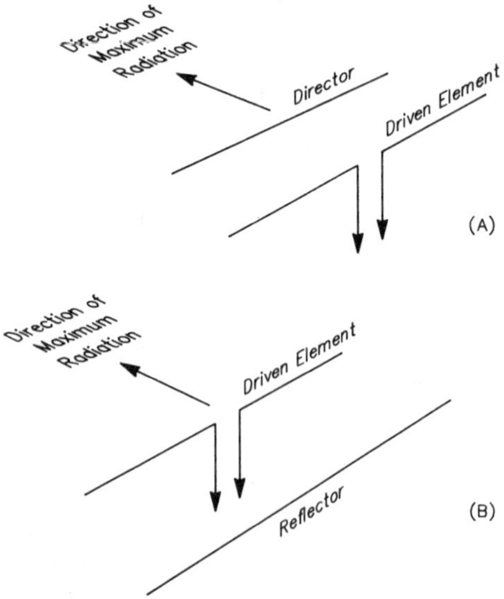

Figure 9-7—In a directional antenna, the reflector element is placed behind the driven element. The director goes in front of the driven element.

antenna. See Figure 9-7. The direction of maximum radiation from a parasitic beam antenna is from the reflector through the driven element to the director. The term *major lobe* refers to the region of maximum radiation from a directional antenna. The major lobe is also sometimes referred to as the *main lobe*. Communication in different directions may be achieved by rotating the array in the *azimuthal*, or *horizontal, plane*, to point it in different compass directions.

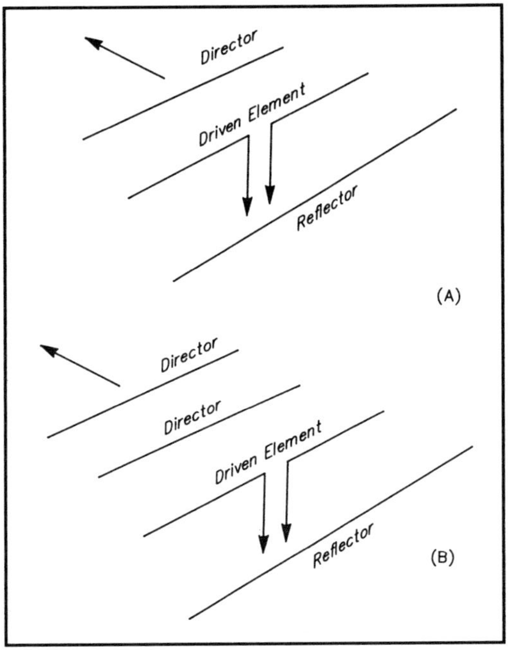

Figure 9-8—At A, a three-element Yagi. At B, a four-element Yagi with two directors.

Yagi Arrays

The **Yagi antennas** in Figure 9-8 are examples of antennas that make use of parasitic elements to produce a *unidirectional radiation pattern*. (This means most of the radio signal goes in one direction.) A Yagi antenna has at least two elements. One element is a driven element. The other elements are directors and/or reflectors. These elements are usually parallel to each other and made of straight metal tubing. They line up with each other along the antenna *boom*.

Though typical HF antennas of this type have three elements, some may have six or more. Multiband Yagi antennas have many elements. Some elements work on some frequencies and others are used for different frequencies. The radiation pattern of the Yagi antenna is shown in Figure 9-9. This pattern indicates that the antenna will reject signals coming from the sides and back. It selects mainly those signals from a desired direction.

The length of the driven element in the most common type of Yagi antenna is approximately an electrical half-wavelength. In a three-element beam, the director is approximately 5% shorter than the driven element. The reflector is approximately 5% longer than the driven element. These lengths can vary considerably, however. The actual lengths depend on the spacing between elements and the diameter of the elements. Whether the elements are made from tapered or cylindrical tubing also makes a difference.

Cubical Quad Antennas

The **cubical quad antenna** also uses parasitic elements. This antenna is sometimes simply called a *quad*. The two elements of the quad antenna are usually wire loops. The total length of the wire in the driven element is approximately one electrical wavelength. A typical quad, shown in Figure 9-10, has two elements—a *driven element* and a *reflector*. A two-element quad could also use a driven element with a *director*. You can add more elements, such as a reflector and one or more directors.

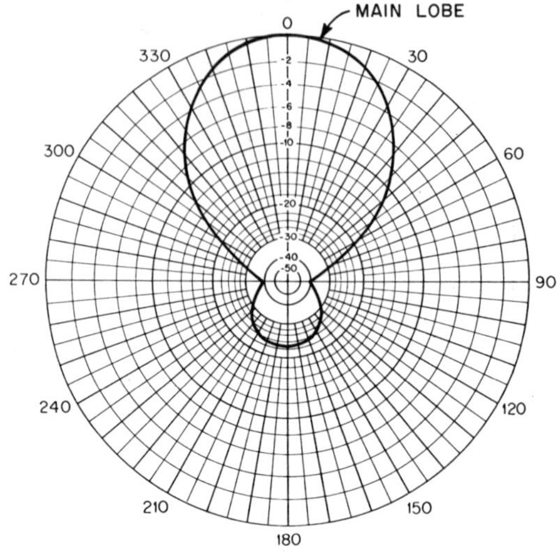

Figure 9-9—The directive pattern for a typical three-element Yagi. Note that this pattern is essentially unidirectional, with most of the radiation in the direction of the main lobe.

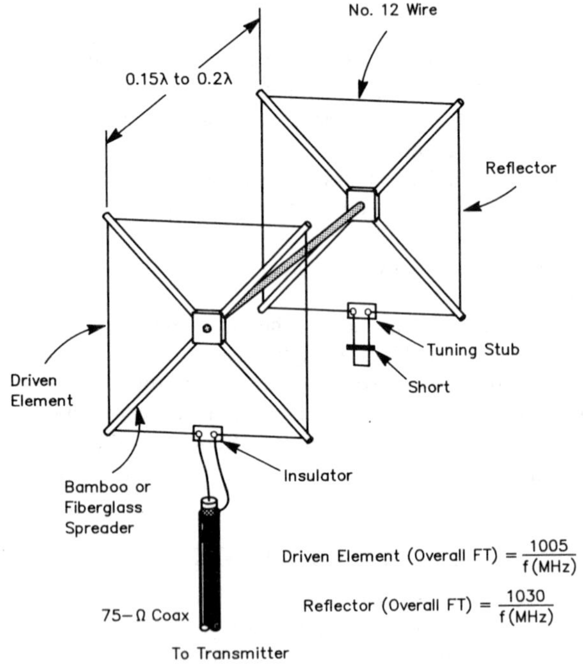

Figure 9-10—The cubical quad antenna. The total length of the driven element is about one wavelength. This antenna can be fed directly with coaxial cable.

The radiation pattern of a typical quad is similar to that of the Yagi shown in Figure 9-9.

The elements of the quad are usually square. Each loop is about an electrical wavelength long. Each side of the square is about ¼-λ long.

Delta Loop Antennas

The **delta loop antenna**, shown in Figure 9-11, is a variation on the quad. A delta loop antenna has triangular elements, rather than square. The total loop length is still one wavelength. Divide the total length by 3 to find the length of each side of the elements. The radiation pattern of a delta loop is similar to that of the quad and Yagi, shown in Figure 9-9.

[Now turn to Chapter 10 and study questions T9A01 through T9A08. Review this section if any of these questions give you difficulty.]

POLARIZATION OF ANTENNAS AND RADIO WAVES

An electromagnetic wave consists of moving electric and magnetic fields. Remember that a field is an invisible force of nature. We can't see radio waves, but we can show a representation of where the energy is in the electric and magnetic fields.

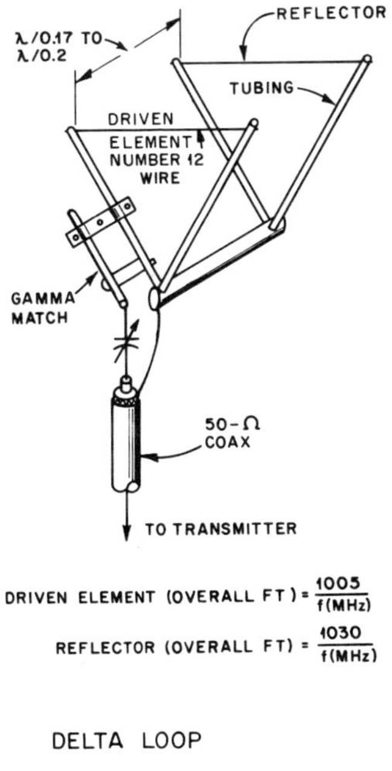

Figure 9-11—A delta loop antenna. The total length of each element can be found using the same equations as used for a quad antenna. The antenna is fed with a gamma match.

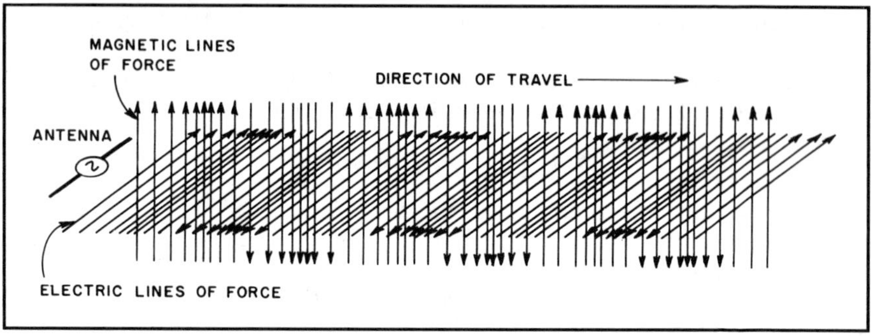

Figure 9-12—A horizontally polarized electromagnetic wave. The electrical lines of force are parallel to the ground. The fields are stronger where the arrows are closer together. The points of maximum field strength correspond to the points of maximum amplitude for the voltage producing the RF signal.

We can visualize a traveling radio wave as looking something like Figure 9-12. The lines of electric and magnetic force are at right angles (90°) to each other. They are also perpendicular (90°) to the direction of travel. These fields can have any position with respect to the earth.

Horizontal Polarization

Polarization is defined by the direction of the electric lines of force in a radio wave. A wave with electric lines of force that are parallel to the surface of the earth is a **horizontally polarized wave**.

The polarization of a radio wave as it leaves the antenna is determined by the orientation of the antenna. For example, a half-wavelength dipole parallel to the surface of the earth transmits a horizontally polarized wave.

Vertical Polarization

The electric lines of force can also be perpendicular to the earth. A wave with vertical electric field lines is a **vertically polarized wave**. A half-wavelength dipole perpendicular to the surface of the earth transmits a vertically polarized wave. A mobile whip antenna is mounted vertically on a car. It transmits a wave that is vertically polarized.

The quarter-wavelength vertical antenna is a popular HF antenna because is provides low-angle radiation when a beam or dipole cannot be placed far enough above ground. Low-angle radiation refers to signals that travel closer to the horizon, rather than signals that are high above the horizon. Low-angle radiation is usually better when you are trying to contact distant stations. Vertical antennas of any length radiate vertically polarized waves.

Polarization is most important when installing antennas for VHF or UHF. Propagation at these frequencies is mostly line of sight. The polarization of a VHF or UHF signal does not change from transmitting antenna to receiving antenna. Best signal reception occurs when both transmitting and receiving stations use the same polarization. The polarization of an HF signal may change many times as it passes through the ionosphere. Antenna polarization at HF is not as important.

Most VHF/UHF FM and data communications is done with vertically polarized

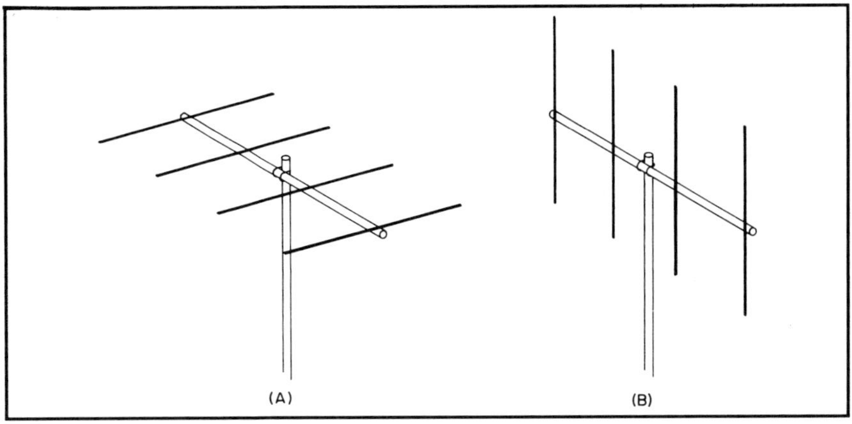

Figure 9-13—The orientation of the elements in a Yagi antenna determines the polarization of the transmitted wave. If the elements are horizontal as shown at A, the wave has horizontal polarization. Vertical mounting, shown at B, produces a vertically polarized wave.

antennas. Vertically polarized antennas are more popular for VHF/UHF FM operation, because the antennas used on cars are almost always verticals. Vertical antennas are more useful for repeaters and home-station use on these bands as well, because they are not directional. For long-distance FM work, a vertically polarized beam is the best antenna. Data communication on the VHF/UHF bands is also mostly done with vertically polarized antennas. VHF/UHF CW and SSB operation, however, is done mostly with horizontally polarized antennas.

Yagi beam antennas can have either horizontal or vertical polarization. If the antenna elements are parallel to the earth, or horizontal, the antenna will produce horizontally polarized waves.

If the elements are turned to be perpendicular to the earth, the antenna will produce vertically polarized waves. Figure 9-13 shows two Yagi antennas. The antenna at A produces horizontally polarized waves and the one at B produces vertically polarized waves. Notice that the boom is horizontal in both cases.

Most man-made noise tends to be vertically polarized. Thus, a horizontally polarized antenna will receive less noise of this type than a vertical antenna will.

[Turn to Chapter 10 and study questions T9B01 through T9B05. Review the material in this section as needed.]

RF SAFETY

No matter what antenna type you choose to build, you should remember a few key points about safety. If possible, ask someone with experience to help you plan and carry out your installation.

If you build a wire antenna, be sure the antenna is out of reach of passersby. The ends should be high enough that a person on the ground cannot touch them. Even at low power levels, enough voltage exists at the antenna ends when you transmit to cause **RF burns**. If you have a vertical antenna with its base at ground level, you might consider building a wooden safety fence around it. The fence should be at least 4 feet away from the antenna. Do not use a metal fence, as this will interfere with the proper antenna operation. If you have a tower, you should either fence it or put shields directly on the tower to prevent unauthorized climbing.

If you use a parallel-conductor feed line, such as open-wire line, you should use care to install the feed line so no one can touch the wires when you are transmitting. The voltage on this line could cause RF burns.

Be especially certain that your antenna is not close to any power lines. That is the only way you can be sure it won't come in contact with them!

When using a hand-held transceiver that runs more than a few watts, be careful! Always keep the antenna away from your head and away from others standing nearby. Some antennas are safer than others. A short, helically wound flexible antenna (called a "rubber duck") is not the safest kind. It concentrates its radiation in a small area near your head. A longer antenna, such as a ½-λ whip, would be safer for a hand-held. A ½-λ antenna concentrates its radiation at its center, which will be farther from your head.

[Congratulations. This completes your study of antennas for your Technician exam. You will probably want to learn more about antennas later. After you pass the exam and have your license you'll be able to experiment with lots of antenna types. Now turn to Chapter 10 and study questions T9C09 and T9C10. Review this section if these questions give you difficulty.]

10 Technician (Element 3A) Question Pool— With Answers

DON'T START HERE!

This chapter contains the complete question pool for the Element 3A exam. Element 3A is part of the Technician license exam. You have credit for passing the Element 2 exam because you have a Novice license. If you do not have a Novice license you will have to pass that exam. *In that case, you should be studying a copy of the ARRL publication* Now You're Talking!

Before you read the questions and answers printed in this chapter, be sure to read the text in the previous chapters. Use these questions as review exercises, when the text tells you to study them. Don't try to memorize all the questions and answers.

The material presented in this book has been carefully written and presented to guide you step-by-step through the learning process. By understanding the electronics principles and Amateur Radio concepts as they are presented, your insight into our hobby and your appreciation for the privileges granted by an Amateur Radio license will be greatly enhanced.

This pool, released by the Volunteer Examiner Coordinators' Question Pool Committee, will be used on exams beginning July 1, 1993. The pool is scheduled to be used until June 30, 1997. Changes to FCC Rules and other factors may result in earlier revisions. Such changes will be announced in *QST* and other Amateur Radio publications.

How Many Questions?

The FCC specifies that an Element 3A exam must include 25 questions. Nine subelements make up each exam question pool. The FCC also specifies the number of questions from each subelement that must appear on the exam. For example, there must be three questions from the Operating Procedures subelement and two questions from the Circuit Components subelement. Table 10-1 summarizes the number of questions to be selected from each subelement on the exam. These numbers are also included at the beginning of each subelement in the question pool.

Table 10-1

Technician Exam Content

Subelement	*Topic*	*Number of questions*
T1	Commission's Rules	5
T2	Operating Procedures	3
T3	Radio-Wave Propagation	3
T4	Amateur Radio Practices	4
T5	Electrical Principles	2
T6	Circuit Components	2
T7	Practical Circuits	1
T8	Signals and Emissions	2
T9	Antennas and Feed Lines	3

The Volunteer Examiner Coordinators' Question Pool Committee has broken the subelements into smaller groups. Each subelement has the same number of groups as there are questions from that subelement on the exam. This means there are three groups for the Operating Procedures subelement and two groups for the Circuit Components subelement. The Question Pool Committee suggests one question from each group be used on the exam. These groups are called out with a list of topics given in bold type in the question pool. They are labeled alphabetically within each subelement. The five groups in the Commission's Rules subelement are labeled T1A through T1E.

The question numbers used in the question pool relate to the syllabus or study guide printed at the end of Chapter 1. The syllabus is an outline of topics covered by the exam. Each question number begins with a T to indicate the question is from the Technician question pool. Next is a number to indicate which of the nine subelements the question is from. A letter following this number indicates which subelement group the question is from. Each question number ends with a two-digit number to specify its position in the set. So question T2A01 is the first question in the A group of the second subelement. Question T9C08 is the eighth question in the C group of the ninth subelement.

This question pool contains more than ten times the number of questions necessary to make up an exam. This ensures that the examiners have sufficient questions to choose from when they make up an exam.

Who Picks the Questions?

The FCC allows Volunteer-Examiner teams to select the questions that will be used on amateur exams. If your test is coordinated by the ARRL/VEC, however, your test will be prepared by the VEC, or using a computer program supplied by the VEC.

Question Pool Format

The rest of this chapter contains the entire Element 3A question pool. The answer key is printed on pages 10-56 and 10-57. (Paper clips make excellent place markers to help you find your place in the text, the question pool and the answer key as you study.) After making your best effort to answer the questions, you can look at the answers to check your understanding.

Along with the answers, we have included references to the section of Part 97 that each of the "Commission's Rules" questions of Subelement 1 refer to. *The FCC Rule Book*, published by ARRL includes the complete text for Part 97, along with text to explain the rules. We recommend you use that book as a reference for the answers to these questions.

We have included page numbers along with the answers to the other questions. These page numbers indicate where in this book you will find the text discussion related to each question. If you have any problems with a question, refer to the page listed for that question. You may have to study beyond the listed page number to review all the related material.

The Question Pool Committee included all the drawings for the question pool on a single page. We placed a copy of this page at the beginning of the question pool, for your information. You should expect to receive a page like this as part of your exam papers. Notice there is a representation of the Ohm's Law Circle you learned about in Chapter 5 on the page. This will help you remember the Ohm's Law equations for your exam. You may not need this equation for your exam, but it will be available for you on the Figure sheet. For your convenience, we also placed the individual figures near the questions that refer to them in the question pool.

Good luck with your studies.

Element 3A (Technician)

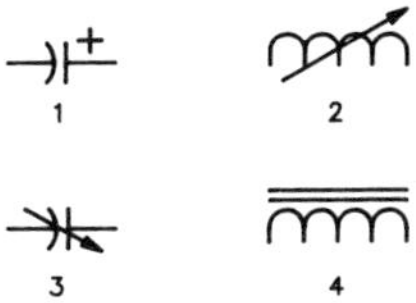

Figure T6-1

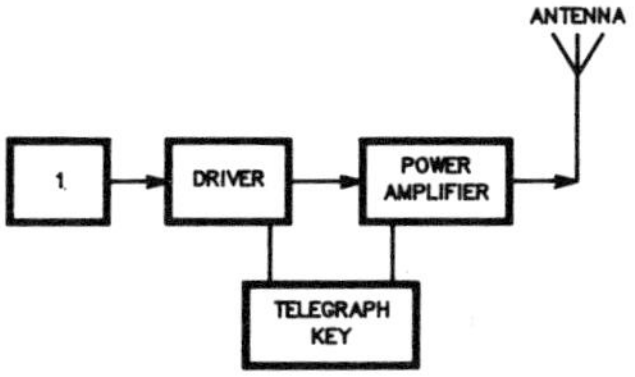

Figure T7-1

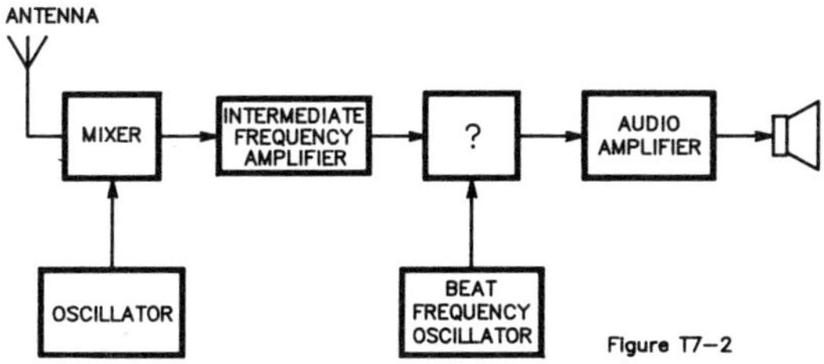

Figure T7-2

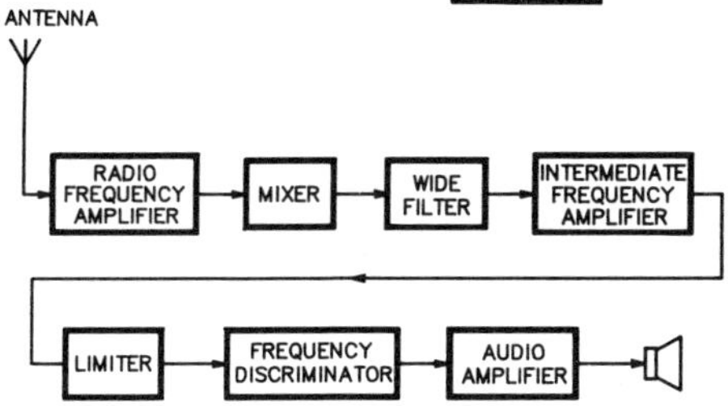

Figure T7-3

$$\frac{E}{I \quad R}$$

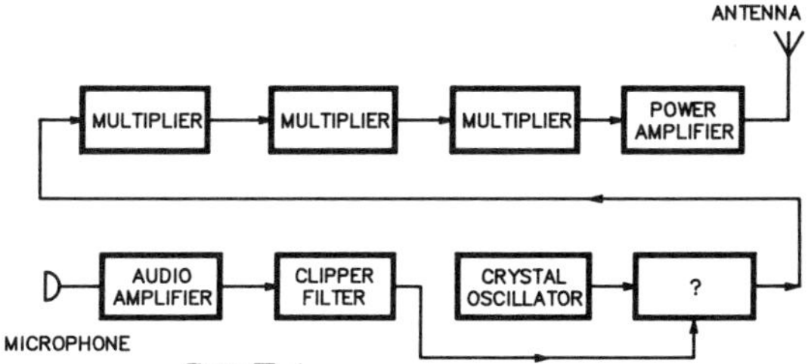

Figure T7-4

As Prepared by the
Question Pool Committee
for all examinations
administered
July 1, 1993 through June 30, 1997

Element 3A (Technician Class) Question Pool

SUBELEMENT T1—COMMISSION'S RULES
[5 exam questions—5 groups]

T1A Station control, frequency privileges authorized to the Technician-class control operator, term of licenses, grace periods and modifications of licenses.

T1A01
What is the control point of an amateur station?
A. The on/off switch of the transmitter
B. The input/output port of a packet controller
C. The variable frequency oscillator of a transmitter
D. The location at which the control operator function is performed

T1A02
What is the term for the location at which the control operator function is performed?
A. The operating desk
B. The control point
C. The station location
D. The manual control location

T1A03
What must you do to renew or change your operator/primary station license?
A. Properly fill out FCC Form 610 and send it to the FCC in Gettysburg, PA
B. Properly fill out FCC Form 610 and send it to the nearest FCC field office
C. Properly fill out FCC Form 610 and send it to the FCC in Washington, DC
D. An amateur license never needs changing or renewing

T1A04
What is the "grace period" during which the FCC will renew an expired 10-year license?
A. 2 years
B. 5 years
C. 10 years
D. There is no grace period

T1A05
Which of the following frequencies may a Technician operator who has passed a Morse code test use?
A. 7.1 - 7.2 MHz
B. 14.1 - 14.2 MHz
C. 21.1 - 21.2 MHz
D. 28.1 - 29.2 MHz

T1A06
Which operator licenses authorize privileges on 52.525 MHz?
A. Extra, Advanced only
B. Extra, Advanced, General only
C. Extra, Advanced, General, Technician only
D. Extra, Advanced, General, Technician, Novice

T1A07
Which operator licenses authorize privileges on 146.52 MHz?
A. Extra, Advanced, General, Technician, Novice
B. Extra, Advanced, General, Technician only
C. Extra, Advanced, General only
D. Extra, Advanced only

T1A08
Which operator licenses authorize privileges on 223.50 MHz?
A. Extra, Advanced, General, Technician, Novice
B. Extra, Advanced, General, Technician only
C. Extra, Advanced, General only
D. Extra, Advanced only

T1A09
Which operator licenses authorize privileges on 446.0 MHz?
A. Extra, Advanced, General, Technician, Novice
B. Extra, Advanced, General, Technician only
C. Extra, Advanced, General only
D. Extra, Advanced only

T1A10
In addition to passing the Technician written examination (Elements 2 and 3A), what must you do before you are allowed to use amateur frequencies below 30 MHz?
A. Nothing special is needed; all Technicians may use the HF bands at any time
B. You must notify the FCC that you intend to operate on the HF bands
C. You must attend a class to learn about HF communications
D. You must pass a Morse code test (either Element 1A, 1B or 1C)

T1A11
If you are a Technician licensee, what must you have to prove that you are authorized to use the Novice amateur frequencies below 30 MHz?
A. A certificate from the FCC showing that you have notified them that you will be using the HF bands
B. A certificate from an instructor showing that you have attended a class in HF communications
C. Written proof of having passed a Morse code test
D. No special proof is required before using the HF bands

T1B Emission privileges for Technician-class control operator, frequency selection and sharing, transmitter power.

T1B01
At what point in your station is transceiver power measured?
A. At the power supply terminals inside the transmitter or amplifier
B. At the final amplifier input terminals inside the transmitter or amplifier
C. At the antenna terminals of the transmitter or amplifier
D. On the antenna itself, after the feed line

T1B02
What is the term for the average power supplied to an antenna transmission line during one RF cycle at the crest of the modulation envelope?
A. Peak transmitter power
B. Peak output power
C. Average radio-frequency power
D. Peak envelope power

T1B03
What is the maximum transmitting power permitted an amateur station in beacon operation?
A. 10 watts PEP output
B. 100 watts PEP output
C. 500 watts PEP output
D. 1500 watts PEP output

T1B04
If the FCC rules say that the amateur service is a secondary user of a frequency band, and another service is a primary user, what does this mean?
A. Nothing special; all users of a frequency band have equal rights to operate
B. Amateurs are only allowed to use the frequency band during emergencies
C. Amateurs are allowed to use the frequency band only if they do not cause harmful interference to primary users
D. Amateurs must increase transmitter power to overcome any interference caused by primary users

T1B05
If you are using a frequency within a band assigned to the amateur service on a secondary basis, and a station assigned to the primary service on that band causes interference, what action should you take?
A. Notify the FCC's regional Engineer in Charge of the interference
B. Increase your transmitter's power to overcome the interference
C. Attempt to contact the station and request that it stop the interference
D. Change frequencies; you may be causing harmful interference to the other station, in violation of FCC rules

T1B06
What rule applies if two amateur stations want to use the same frequency?
A. The station operator with a lesser class of license must yield the frequency to a higher-class licensee
B. The station operator with a lower power output must yield the frequency to the station with a higher power output
C. Both station operators have an equal right to operate on the frequency
D. Station operators in ITU Regions 1 and 3 must yield the frequency to stations in ITU Region 2

T1B07
What emission type may always be used for station identification, regardless of the transmitting frequency?
A. CW
B. RTTY
C. MCW
D. Phone

T1B08
On what frequencies within the 6-meter band may phone emissions be transmitted?
A. 50.0 - 54.0 MHz only
B. 50.1 - 54.0 MHz only
C. 51.0 - 54.0 MHz only
D. 52.0 - 54.0 MHz only

T1B09
On what frequencies within the 2-meter band may image emissions be transmitted?
A. 144.1 - 148.0 MHz only
B. 146.0 - 148.0 MHz only
C. 144.0 - 148.0 MHz only
D. 146.0 - 147.0 MHz only

T1B10
What is the maximum transmitting power permitted an amateur station on 146.52 MHz?
A. 200 watts PEP output
B. 500 watts ERP
C. 1000 watts DC input
D. 1500 watts PEP output

T1B11
Which band may NOT be used by Earth stations for satellite communications?
A. 6 meters
B. 2 meters
C. 70 centimeters
D. 23 centimeters

T1C Digital communications, station identification, ID with CSCE.

T1C01
If you are a Novice licensee with a Certificate of Successful Completion of Examination (CSCE) for Technician privileges, how do you identify your station when transmitting on 146.34 MHz?
A. You must give your call sign, followed by any suitable word that denotes the slant mark and the identifier "KT"
B. You may not operate on 146.34 until your new license arrives
C. No special form of identification is needed
D. You must give your call sign and the location of the VE examination where you obtained the CSCE

T1C02
What is the maximum frequency shift permitted for RTTY or data transmissions below 50 MHz?
A. 0.1 kHz
B. 0.5 kHz
C. 1 kHz
D. 5 kHz

T1C03
What is the maximum frequency shift permitted for RTTY or data transmissions above 50 MHz?
A. 0.1 kHz or the sending speed, in bauds, whichever is greater
B. 0.5 kHz or the sending speed, in bauds, whichever is greater
C. 5 kHz or the sending speed, in bauds, whichever is greater
D. The FCC rules do not specify a maximum frequency shift above 50 MHz

T1C04
What is the maximum symbol rate permitted for packet transmissions on the 10-meter band?
A. 300 bauds
B. 1200 bauds
C. 19.6 kilobauds
D. 56 kilobauds

T1C05
What is the maximum symbol rate permitted for packet transmissions on the 2-meter band?
A. 300 bauds
B. 1200 bauds
C. 19.6 kilobauds
D. 56 kilobauds

T1C06
What is the maximum symbol rate permitted for RTTY or data transmissions between 28 and 50 MHz?
A. 56 kilobauds
B. 19.6 kilobauds
C. 1200 bauds
D. 300 bauds

T1C07
What is the maximum symbol rate permitted for RTTY or data transmissions between 50 and 222 MHz?
A. 56 kilobauds
B. 19.6 kilobauds
C. 1200 bauds
D. 300 bauds

T1C08
What is the maximum authorized bandwidth of RTTY, data or multiplexed emissions using an unspecified digital code within the frequency range of 50 to 222 MHz?
A. 20 kHz
B. 50 kHz
C. The total bandwidth shall not exceed that of a single-sideband phone emission
D. The total bandwidth shall not exceed 10 times that of a CW emission

T1C09
What is the maximum symbol rate permitted for RTTY or data transmissions above 222 MHz?
A. 300 bauds
B. 1200 bauds
C. 19.6 kilobauds
D. 56 kilobauds

T1C10
What is the maximum authorized bandwidth of RTTY, data or multiplexed emissions using an unspecified digital code within the frequency range of 222 to 450 MHz?
A. 50 kHz
B. 100 kHz
C. 150 kHz
D. 200 kHz

T1C11
What is the maximum authorized bandwidth of RTTY, data or multiplexed emissions using an unspecified digital code within the 70-cm amateur band?
A. 300 kHz
B. 200 kHz
C. 100 kHz
D. 50 kHz

T1D Correct language, phonetics, beacons and radio control of model craft and vehicles.

T1D01
What is an amateur station called which transmits communications for the purpose of observation of propagation and reception?
A. A beacon
B. A repeater
C. An auxiliary station
D. A radio control station

T1D02
What is the fastest code speed a repeater may use for automatic identification?
A. 13 words per minute
B. 20 words per minute
C. 25 words per minute
D. There is no limitation

T1D03
If you are using a language besides English to make a contact, what language must you use when identifying your station?
A. The language being used for the contact
B. The language being used for the contact, providing the US has a third-party communications agreement with that country
C. English
D. Any language of a country which is a member of the International Telecommunication Union

T1D04
What do the FCC rules suggest you use as an aid for correct station identification when using phone?
A. A speech compressor
B. Q signals
C. A phonetic alphabet
D. Unique words of your choice

T1D05
What minimum class of amateur license must you hold to operate a beacon station?
A. Novice
B. Technician
C. General
D. Amateur Extra

T1D06
If a repeater is causing harmful interference to another repeater and a frequency coordinator has recommended the operation of one station only, who is responsible for resolving the interference?
A. The licensee of the unrecommended repeater
B. Both repeater licensees
C. The licensee of the recommended repeater
D. The frequency coordinator

T1D07
If a repeater is causing harmful interference to another amateur repeater and a frequency coordinator has recommended the operation of both stations, who is responsible for resolving the interference?
A. The licensee of the repeater which has been recommended for the longest period of time
B. The licensee of the repeater which has been recommended the most recently
C. The frequency coordinator
D. Both repeater licensees

T1D08
If a repeater is causing harmful interference to another repeater and a frequency coordinator has NOT recommended either station, who is primarily responsible for resolving the interference?
A. Both repeater licensees
B. The licensee of the repeater which has been in operation for the longest period of time
C. The licensee of the repeater which has been in operation for the shortest period of time
D. The frequency coordinator

T1D09
What minimum information must be on a label affixed to a transmitter used for telecommand (control) of model craft?
A. Station call sign
B. Station call sign and the station licensee's name
C. Station call sign and the station licensee's name and address
D. Station call sign and the station licensee's class of license

T1D10
What are the station identification requirements for an amateur transmitter used for telecommand (control) of model craft?
A. Once every ten minutes
B. Once every ten minutes, and at the beginning and end of each transmission
C. At the beginning and end of each transmission
D. Station identification is not required if the transmitter is labeled with the station licensee's name, address and call sign

T1D11
What is the maximum transmitter power an amateur station is allowed when used for telecommand (control) of model craft?
A. One milliwatt
B. One watt
C. Two watts
D. Three watts

T1E Emergency communications; broadcasting; permissible one-way, satellite and third-party communication; indecent and profane language.

T1E01
What is meant by the term broadcasting?
A. Transmissions intended for reception by the general public, either direct or relayed
B. Retransmission by automatic means of programs or signals from non-amateur stations
C. One-way radio communications, regardless of purpose or content
D. One-way or two-way radio communications between two or more stations

T1E02
Which of the following one-way communications may not be transmitted in the amateur service?
A. Telecommands to model craft
B. Broadcasts intended for the general public
C. Brief transmissions to make adjustments to the station
D. Morse code practice

T1E03
What kind of payment is allowed for third-party messages sent by an amateur station?
A. Any amount agreed upon in advance
B. Donation of equipment repairs
C. Donation of amateur equipment
D. No payment of any kind is allowed

T1E04
When may you send obscene words from your amateur station?
A. Only when they do not cause interference to other communications
B. Never; obscene words are prohibited in amateur transmissions
C. Only when they are not retransmitted through a repeater
D. Any time, but there is an unwritten rule among amateurs that they should not be used on the air

T1E05
When may you send indecent words from your amateur station?
A. Only when they do not cause interference to other communications
B. Only when they are not retransmitted through a repeater
C. Any time, but there is an unwritten rule among amateurs that they should not be used on the air
D. Never; indecent words are prohibited in amateur transmissions

T1E06
When may you send profane words from your amateur station?
A. Only when they do not cause interference to other communications
B. Only when they are not retransmitted through a repeater
C. Never; profane words are prohibited in amateur transmissions
D. Any time, but there is an unwritten rule among amateurs that they should not be used on the air

T1E07
If you wanted to use your amateur station to retransmit communications between a space shuttle and its associated Earth stations, what agency must first give its approval?
A. The FCC in Washington, DC
B. The office of your local FCC Engineer In Charge (EIC)
C. The National Aeronautics and Space Administration
D. The Department of Defense

T1E08
When are third-party messages allowed to be sent to a foreign country?
A. When sent by agreement of both control operators
B. When the third party speaks to a relative
C. They are not allowed under any circumstances
D. When the US has a third-party agreement with the foreign country or the third party is qualified to be a control operator

T1E09
If you let an unlicensed third party use your amateur station, what must you do at your station's control point?
A. You must continuously monitor and supervise the third-party's participation
B. You must monitor and supervise the communication only if contacts are made in countries which have no third-party communications agreement with the US
C. You must monitor and supervise the communication only if contacts are made on frequencies below 30 MHz
D. You must key the transmitter and make the station identification

T1E10
If a disaster disrupts normal communication systems in an area where the amateur service is regulated by the FCC, what kinds of transmissions may stations make?
A. Those which are necessary to meet essential communication needs and facilitate relief actions
B. Those which allow a commercial business to continue to operate in the affected area
C. Those for which material compensation has been paid to the amateur operator for delivery into the affected area
D. Those which are to be used for program production or newsgathering for broadcasting purposes

T1E11
What information is included in an FCC declaration of a temporary state of communication emergency?
A. A list of organizations authorized to use radio communications in the affected area
B. A list of amateur frequency bands to be used in the affected area
C. Any special conditions and special rules to be observed during the emergency
D. An operating schedule for authorized amateur emergency stations

SUBELEMENT T2—OPERATING PROCEDURES
[3 exam questions—3 groups]

T2A Repeater operation, courteous operation.

T2A01
How do you call another station on a repeater if you know the station's call sign?
A. Say "break, break 79," then say the station's call sign
B. Say the station's call sign, then identify your own station
C. Say "CQ" three times, then say the station's call sign
D. Wait for the station to call "CQ," then answer it

T2A02
Why should you pause briefly between transmissions when using a repeater?
A. To check the SWR of the repeater
B. To reach for pencil and paper for third-party communications
C. To listen for anyone wanting to break in
D. To dial up the repeater's autopatch

T2A03
Why should you keep transmissions short when using a repeater?
A. A long transmission may prevent someone with an emergency from using the repeater
B. To see if the receiving station operator is still awake
C. To give any listening non-hams a chance to respond
D. To keep long-distance charges down

T2A04
What is the proper way to break into a conversation on a repeater?
A. Wait for the end of a transmission and start calling the desired party
B. Shout, "break, break!" to show that you're eager to join the conversation
C. Turn on an amplifier and override whoever is talking
D. Say your call sign during a break between transmissions

T2A05
What is the purpose of repeater operation?
A. To cut your power bill by using someone else's higher power system
B. To help mobile and low-power stations extend their usable range
C. To transmit signals for observing propagation and reception
D. To make calls to stores more than 50 miles away

T2A06
What causes a repeater to "time out"?
A. The repeater's battery supply runs out
B. Someone's transmission goes on longer than the repeater allows
C. The repeater gets too hot and stops transmitting until its circuitry cools off
D. Something is wrong with the repeater

T2A07
During commuting rush hours, which type of repeater operation should be discouraged?
A. Mobile stations
B. Low-power stations
C. Highway traffic information nets
D. Third-party communications nets

T2A08
What is a courtesy tone (used in repeater operations)?
A. A sound used to identify the repeater
B. A sound used to indicate when a transmission is complete
C. A sound used to indicate that a message is waiting for someone
D. A sound used to activate a receiver in case of severe weather

T2A09
What is the meaning of: "Your signal is full quieting..."?
A. Your signal is strong enough to overcome all receiver noise
B. Your signal has no spurious sounds
C. Your signal is not strong enough to be received
D. Your signal is being received, but no audio is being heard

T2A10
How should you give a signal report over a repeater?
A. Say what your receiver's S-meter reads
B. Always say: "Your signal report is five five..."
C. Say the amount of signal quieting into the repeater
D. Try to imitate the sound quality you are receiving

T2A11
What is a repeater called which is available for anyone to use?
A. An open repeater
B. A closed repeater
C. An autopatch repeater
D. A private repeater

T2A12
What is the usual input/output frequency separation for repeaters in the 2-meter band?
A. 600 kHz
B. 1.0 MHz
C. 1.6 MHz
D. 5.0 MHz

T2A13
What is the usual input/output frequency separation for repeaters in the 1.25-meter band?
A. 600 kHz
B. 1.0 MHz
C. 1.6 MHz
D. 5.0 MHz

T2A14
What is the usual input/output frequency separation for repeaters in the 70-centimeter band?
A. 600 kHz
B. 1.0 MHz
C. 1.6 MHz
D. 5.0 MHz

T2A15
Why should local amateur communications use VHF and UHF frequencies instead of HF frequencies?
A. To minimize interference on HF bands capable of long-distance communication
B. Because greater output power is permitted on VHF and UHF
C. Because HF transmissions are not propagated locally
D. Because signals are louder on VHF and UHF frequencies

T2A16
How might you join a closed repeater system?
A. Contact the control operator and ask to join
B. Use the repeater until told not to
C. Use simplex on the repeater input until told not to
D. Write the FCC and report the closed condition

T2A17
How can on-the-air interference be minimized during a lengthy transmitter testing or loading-up procedure?
A. Choose an unoccupied frequency
B. Use a dummy load
C. Use a non-resonant antenna
D. Use a resonant antenna that requires no loading-up procedure

T2A18
What is the proper way to ask someone their location when using a repeater?
A. What is your QTH
B. What is your 20
C. Where are you
D. Locations are not normally told by radio

T2B Simplex operations, Q signals, RST signal reporting, repeater frequency coordination.

T2B01
Why should simplex be used where possible, instead of using a repeater?
A. Signal range will be increased
B. Long distance toll charges will be avoided
C. The repeater will not be tied up unnecessarily
D. Your antenna's effectiveness will be better tested

T2B02
If you are talking to a station using a repeater, how would you find out if you could communicate using simplex instead?
A. See if you can clearly receive the station on the repeater's input frequency
B. See if you can clearly receive the station on a lower frequency band
C. See if you can clearly receive a more distant repeater
D. See if a third station can clearly receive both of you

T2B03
If you are operating simplex on a repeater frequency, why would it be good amateur practice to change to another frequency?
A. The repeater's output power may ruin your station's receiver
B. There are more repeater operators than simplex operators
C. Changing the repeater's frequency is not practical
D. Changing the repeater's frequency requires the authorization of the FCC

T2B04
What is a repeater frequency coordinator?
A. Someone who organizes the assembly of a repeater station
B. Someone who provides advice on what kind of repeater to buy
C. The person whose call sign is used for a repeater's identification
D. A person or group that recommends frequencies for repeater usage

T2B05
What is the proper Q signal to use to see if a frequency is in use before transmitting on CW?
A. QRV?
B. QRU?
C. QRL?
D. QRZ?

T2B06
What is one meaning of the Q signal "QSY"?
A. Change frequency
B. Send more slowly
C. Send faster
D. Use more power

T2B07
What is one meaning of the Q signal "QSO"?
A. A contact is confirmed
B. A conversation is in progress
C. A contact is ending
D. A conversation is desired

T2B08
What is the proper Q signal to use to ask if someone is calling you on CW?
A. QSL?
B. QRZ?
C. QRL?
D. QRT?

T2B09
What is the meaning of: "Your signal report is five seven..."?
A. Your signal is perfectly readable and moderately strong
B. Your signal is perfectly readable, but weak
C. Your signal is readable with considerable difficulty
D. Your signal is perfectly readable with near pure tone

T2B10
What is the meaning of: "Your signal report is three three..."?
A. The contact is serial number thirty-three
B. The station is located at latitude 33 degrees
C. Your signal is readable with considerable difficulty and weak in strength
D. Your signal is unreadable, very weak in strength

T2B11
What is the meaning of: "Your signal report is five nine plus 20 dB..."?
A. Your signal strength has increased by a factor of 100
B. Repeat your transmission on a frequency 20 kHz higher
C. The bandwidth of your signal is 20 decibels above linearity
D. A relative signal-strength meter reading is 20 decibels greater than strength 9

T2C Distress calling and emergency drills and communications—operations and equipment, Radio Amateur Civil Emergency Service (RACES).

T2C01
What is the proper distress call to use when operating phone?
A. Say "MAYDAY" several times
B. Say "HELP" several times
C. Say "EMERGENCY" several times
D. Say "SOS" several times

T2C02
What is the proper distress call to use when operating CW?
A. MAYDAY
B. QRRR
C. QRZ
D. SOS

T2C03
What is the proper way to interrupt a repeater conversation to signal a distress call?
A. Say "BREAK" twice, then your call sign
B. Say "HELP" as many times as it takes to get someone to answer
C. Say "SOS," then your call sign
D. Say "EMERGENCY" three times

T2C04
With what organization must you register before you can participate in RACES drills?
A. A local Amateur Radio club
B. A local racing organization
C. The responsible civil defense organization
D. The Federal Communications Commission

T2C05
What is the maximum number of hours allowed per week for RACES drills?
A. One
B. Six, but not more than one hour per day
C. Eight
D. As many hours as you want

T2C06
How must you identify messages sent during a RACES drill?
A. As emergency messages
B. As amateur traffic
C. As official government messages
D. As drill or test messages

T2C07
What is one reason for using tactical call signs such as "command post" or "weather center" during an emergency?
A. They keep the general public informed about what is going on
B. They are more efficient and help coordinate public-service communications
C. They are required by the FCC
D. They increase goodwill between amateurs

T2C08
What type of messages concerning a person's well-being are sent into or out of a disaster area?
A. Routine traffic
B. Tactical traffic
C. Formal message traffic
D. Health and Welfare traffic

T2C09
What are messages called which are sent into or out of a disaster area concerning the immediate safety of human life?
A. Tactical traffic
B. Emergency traffic
C. Formal message traffic
D. Health and Welfare traffic

T2C10
Why is it a good idea to have a way to operate your amateur station without using commercial AC power lines?
A. So you may use your station while mobile
B. So you may provide communications in an emergency
C. So you may operate in contests where AC power is not allowed
D. So you will comply with the FCC rules

T2C11
What is the most important accessory to have for a hand-held radio in an emergency?
A. An extra antenna
B. A portable amplifier
C. Several sets of charged batteries
D. A microphone headset for hands-free operation

T2C12
Which type of antenna would be a good choice as part of a portable HF amateur station that could be set up in case of an emergency?
A. A three-element quad
B. A three-element Yagi
C. A dipole
D. A parabolic dish

SUBELEMENT T3—RADIO-WAVE PROPAGATION
[3 exam questions—3 groups]

T3A Ionosphere, ionospheric regions, solar radiation.

T3A01
What is the ionosphere?
A. An area of the outer atmosphere where enough ions and free electrons exist to propagate radio waves
B. An area between two air masses of different temperature and humidity, along which radio waves can travel
C. An ionized path in the atmosphere where lightning has struck
D. An area of the atmosphere where weather takes place

T3A02
What is the name of the area that makes long-distance radio communications possible by bending radio waves?
A. Troposphere
B. Stratosphere
C. Magnetosphere
D. Ionosphere

T3A03
What causes the ionosphere to form?
A. Solar radiation ionizing the outer atmosphere
B. Temperature changes ionizing the outer atmosphere
C. Lightning ionizing the outer atmosphere
D. Release of fluorocarbons into the atmosphere

T3A04
What type of solar radiation is most responsible for ionization in the outer atmosphere?
A. Thermal
B. Ionized particle
C. Ultraviolet
D. Microwave

T3A05
Which ionospheric region limits daytime radio communications on the 80-meter band to short distances?
A. D region
B. E region
C. F1 region
D. F2 region

T3A06
Which ionospheric region is closest to the earth?
A. The A region
B. The D region
C. The E region
D. The F region

T3A07
Which ionospheric region most affects sky-wave propagation on the 6-meter band?
A. The D region
B. The E region
C. The F1 region
D. The F2 region

T3A08
Which region of the ionosphere is the least useful for long-distance radio-wave propagation?
A. The D region
B. The E region
C. The F1 region
D. The F2 region

T3A09
Which region of the ionosphere is mainly responsible for long-distance sky-wave radio communications?
A. D region
B. E region
C. F1 region
D. F2 region

T3A10
What two sub-regions of ionosphere exist only in the daytime?
A. Troposphere and stratosphere
B. F1 and F2
C. Electrostatic and electromagnetic
D. D and E

T3A11
Which two daytime ionospheric regions combine into one region at night?
A. E and F1
B. D and E
C. F1 and F2
D. E1 and E2

T3B Ionospheric absorption, causes and variation, maximum usable frequency.

T3B01
Which region of the ionosphere is mainly responsible for absorbing radio signals during the daytime?
A. The F2 region
B. The F1 region
C. The E region
D. The D region

T3B02
When does ionospheric absorption of radio signals occur?
A. When tropospheric ducting occurs
B. When long-wavelength signals enter the D region
C. When signals travel to the F region
D. When a temperature inversion occurs

T3B03
What effect does the D region of the ionosphere have on lower-frequency HF signals in the daytime?
A. It absorbs the signals
B. It bends the radio waves out into space
C. It refracts the radio waves back to earth
D. It has little or no effect on 80-meter radio waves

T3B04
What causes the ionosphere to absorb radio waves?
A. The weather below the ionosphere
B. The ionization of the D region
C. The presence of ionized clouds in the E region
D. The splitting of the F region

T3B05
What is the condition of the ionosphere just before local sunrise?
A. Atmospheric attenuation is at a maximum
B. The D region is above the E region
C. The E region is above the F region
D. Ionization is at a minimum

T3B06
When is the ionosphere most ionized?
A. Dusk
B. Midnight
C. Midday
D. Dawn

T3B07
When is the ionosphere least ionized?
A. Shortly before dawn
B. Just after noon
C. Just after dusk
D. Shortly before midnight

T3B08
When is the E region most ionized?
A. Dawn
B. Midday
C. Dusk
D. Midnight

T3B09
What happens to signals higher in frequency than the critical frequency?
A. They pass through the ionosphere
B. They are absorbed by the ionosphere
C. Their frequency is changed by the ionosphere to be below the maximum usable frequency
D. They are reflected back to their source

T3B10
What causes the maximum usable frequency to vary?
A. The temperature of the ionosphere
B. The speed of the winds in the upper atmosphere
C. The amount of radiation received from the sun, mainly ultraviolet
D. The type of weather just below the ionosphere

T3B11
What does maximum usable frequency mean?
A. The highest frequency signal that will reach its intended destination
B. The lowest frequency signal that will reach its intended destination
C. The highest frequency signal that is most absorbed by the ionosphere
D. The lowest frequency signal that is most absorbed by the ionosphere

T3C Propagation, including ionospheric, tropospheric, line-of-sight scatter propagation, and maximum usable frequency.

T3C01
What kind of propagation would best be used by two stations within each other's skip zone on a certain frequency?
A. Ground-wave
B. Sky-wave
C. Scatter-mode
D. Ducting

T3C02
If you are receiving a weak and distorted signal from a distant station on a frequency close to the maximum usable frequency, what type of propagation is probably occurring?
A. Ducting
B. Line-of-sight
C. Scatter
D. Ground-wave

T3C03
How are VHF signals propagated within the range of the visible horizon?
A. By sky wave
B. By direct wave
C. By plane wave
D. By geometric wave

T3C04
Ducting occurs in which region of the atmosphere?
A. F2
B. Ectosphere
C. Troposphere
D. Stratosphere

T3C05
What effect does tropospheric bending have on 2-meter radio waves?
A. It lets you contact stations farther away
B. It causes them to travel shorter distances
C. It garbles the signal
D. It reverses the sideband of the signal

T3C06
What causes tropospheric ducting of radio waves?
A. A very low pressure area
B. An aurora to the north
C. Lightning between the transmitting and receiving stations
D. A temperature inversion

T3C07
What causes VHF radio waves to be propagated several hundred miles over oceans?
A. A polar air mass
B. A widespread temperature inversion
C. An overcast of cirriform clouds
D. A high-pressure zone

T3C08
In what frequency range does tropospheric ducting most often occur?
A. SW
B. MF
C. HF
D. VHF

T3C09
In what frequency range does sky-wave propagation least often occur?
A. LF
B. MF
C. HF
D. VHF

T3C10
What weather condition may cause tropospheric ducting?
A. A stable high-pressure system
B. An unstable low-pressure system
C. A series of low-pressure waves
D. Periods of heavy rainfall

T3C11
What band conditions might indicate long-range skip on the 6-meter and 2-meter bands?
A. Noise on the 80-meter band
B. The absence of signals on the 10-meter band
C. Very long-range skip on the 10-meter band
D. Strong signals on the 10-meter band from stations about 500-600 miles away

SUBELEMENT T4—AMATEUR RADIO PRACTICES
[4 exam questions—4 groups]

T4A Electrical wiring, including switch location, dangerous voltages and currents.

T4A01
Where should the green wire in a three-wire AC line cord be connected in a power supply?
A. To the fuse
B. To the "hot" side of the power switch
C. To the chassis
D. To the white wire

T4A02
Where should the black (or red) wire in a three-wire AC line cord be connected in a power supply?
A. To the white wire
B. To the green wire
C. To the chassis
D. To the fuse

T4A03
Where should the white wire in a three-wire AC line cord be connected in a power supply?
A. To the side of the power transformer's primary winding that has a fuse
B. To the side of the power transformer's primary winding that does not have a fuse
C. To the chassis
D. To the black wire

T4A04
What document is used by almost every US city as the basis for electrical safety requirements for power wiring and antennas?
A. The Code of Federal Regulations
B. The Proceedings of the IEEE
C. The ITU Radio Regulations
D. The National Electrical Code

T4A05
What document would you use to see if you comply with standard electrical safety rules when building an amateur antenna?
A. The Code of Federal Regulations
B. The Proceedings of the IEEE
C. The National Electrical Code
D. The ITU Radio Regulations

T4A06
Where should fuses be connected on a mobile transceiver's DC power cable?
A. Between the red and black wires
B. In series with just the black wire
C. In series with just the red wire
D. In series with both the red and black wires

T4A07
Why is the retaining screw in one terminal of a wall outlet made of brass while the other one is silver colored?
A. To prevent corrosion
B. To indicate correct wiring polarity
C. To better conduct current
D. To reduce skin effect

T4A08
How much electrical current flowing through the human body is usually fatal?
A. As little as 1/10 of an ampere
B. Approximately 10 amperes
C. More than 20 amperes
D. Current flow through the human body is never fatal

T4A09
Which body organ can be fatally affected by a very small amount of electrical current?
A. The heart
B. The brain
C. The liver
D. The lungs

T4A10
How much electrical current flowing through the human body is usually painful?
A. As little as 1/500 of an ampere
B. Approximately 10 amperes
C. More than 20 amperes
D. Current flow through the human body is never painful

T4A11
What is the minimum voltage which is usually dangerous to humans?
A. 30 volts
B. 100 volts
C. 1000 volts
D. 2000 volts

T4A12
Where should the main power switch for a high-voltage power supply be located?
A. Inside the cabinet, to kill the power if the cabinet is opened
B. On the back side of the cabinet, out of sight
C. Anywhere that can be seen and reached easily
D. A high-voltage power supply should not be switch-operated

T4A13
What precaution should you take when leaning over a power amplifier?
A. Take your shoes off
B. Watch out for loose jewelry contacting high voltage
C. Shield your face from the heat produced by the power supply
D. Watch out for sharp edges which may snag your clothing

T4A14
What is an important safety rule concerning the main electrical box in your home?
A. Make sure the door cannot be opened easily
B. Make sure something is placed in front of the door so no one will be able to get to it easily
C. Make sure others in your home know where it is and how to shut off the electricity
D. Warn others in your home never to touch the switches, even in an emergency

T4A15
What should you do if you discover someone who is being burned by high voltage?
A. Run from the area so you won't be burned too
B. Turn off the power, call for emergency help and give CPR if needed
C. Immediately drag the person away from the high voltage
D. Wait for a few minutes to see if the person can get away from the high voltage on their own, then try to help

T4B Meters, including volt, amp, multi, peak-reading, RF watt and placement, and ratings of fuses and switches.

T4B01
How is a voltmeter usually connected to a circuit under test?
A. In series with the circuit
B. In parallel with the circuit
C. In quadrature with the circuit
D. In phase with the circuit

T4B02
How can the range of a voltmeter be increased?
A. By adding resistance in series with the circuit under test
B. By adding resistance in parallel with the circuit under test
C. By adding resistance in series with the meter, between the meter and the circuit under test
D. By adding resistance in parallel with the meter, between the meter and the circuit under test

T4B03
What happens inside a voltmeter when you switch it from a lower to a higher voltage range?
A. Resistance is added in series with the meter
B. Resistance is added in parallel with the meter
C. Resistance is reduced in series with the meter
D. Resistance is reduced in parallel with the meter

T4B04
How is an ammeter usually connected to a circuit under test?
A. In series with the circuit
B. In parallel with the circuit
C. In quadrature with the circuit
D. In phase with the circuit

T4B05
How can the range of an ammeter be increased?
A. By adding resistance in series with the circuit under test
B. By adding resistance in parallel with the circuit under test
C. By adding resistance in series with the meter
D. By adding resistance in parallel with the meter

T4B06
What does a multimeter measure?
A. SWR and power
B. Resistance, capacitance and inductance
C. Resistance and reactance
D. Voltage, current and resistance

T4B07
Where should an RF wattmeter be connected for the most accurate readings of transmitter output power?
A. At the transmitter output connector
B. At the antenna feed point
C. One-half wavelength from the transmitter output
D. One-half wavelength from the antenna feed point

T4B08
At what line impedance do most RF wattmeters usually operate?
A. 25 ohms
B. 50 ohms
C. 100 ohms
D. 300 ohms

T4B09
What does a directional wattmeter measure?
A. Forward and reflected power
B. The directional pattern of an antenna
C. The energy used by a transmitter
D. Thermal heating in a load resistor

T4B10
If a directional RF wattmeter reads 90 watts forward power and 10 watts reflected power, what is the actual transmitter output power?
A. 10 watts
B. 80 watts
C. 90 watts
D. 100 watts

T4B11
If a directional RF wattmeter reads 96 watts forward power and 4 watts reflected power, what is the actual transmitter output power?
A. 80 watts
B. 88 watts
C. 92 watts
D. 100 watts

T4C Marker generator, crystal calibrator, signal generators and impedance-match indicator.

T4C01
What is a marker generator?
A. A high-stability oscillator that generates reference signals at exact frequency intervals
B. A low-stability oscillator that "sweeps" through a range of frequencies
C. A low-stability oscillator used to inject a signal into a circuit under test
D. A high-stability oscillator which can produce a wide range of frequencies and amplitudes

T4C02
How is a marker generator used?
A. To calibrate the tuning dial on a receiver
B. To calibrate the volume control on a receiver
C. To test the amplitude linearity of a transmitter
D. To test the frequency deviation of a transmitter

T4C03
What device is used to inject a frequency calibration signal into a receiver?
A. A calibrated voltmeter
B. A calibrated oscilloscope
C. A calibrated wavemeter
D. A crystal calibrator

T4C04
What frequency standard may be used to calibrate the tuning dial of a receiver?
A. A calibrated voltmeter
B. Signals from WWV and WWVH
C. A deviation meter
D. A sweep generator

T4C05
How might you check the accuracy of your receiver's tuning dial?
A. Tune to the frequency of a shortwave broadcasting station
B. Tune to a popular amateur net frequency
C. Tune to one of the frequencies of station WWV or WWVH
D. Tune to another amateur station and ask what frequency the operator is using

T4C06
What device produces a stable, low-level signal that can be set to a desired frequency?
A. A wavemeter
B. A reflectometer
C. A signal generator
D. An oscilloscope

T4C07
What is an RF signal generator used for?
A. Measuring RF signal amplitudes
B. Aligning tuned circuits
C. Adjusting transmitter impedance-matching networks
D. Measuring transmission-line impedances

T4C08
What device can measure an impedance mismatch in your antenna system?
A. A field-strength meter
B. An ammeter
C. A wavemeter
D. A reflectometer

T4C09
Where should a reflectometer be connected for best accuracy when reading the impedance match between an antenna and its feed line?
A. At the antenna feed point
B. At the transmitter output connector
C. At the midpoint of the feed line
D. Anywhere along the feed line

T4C10
If you use a 3-30 MHz RF power meter for VHF, how accurate will its readings be?
A. They will not be accurate
B. They will be accurate enough to get by
C. If it properly calibrates to full scale in the set position, they may be accurate
D. They will be accurate providing the readings are multiplied by 4.5

T4C11
If you use a 3-30 MHz SWR meter for VHF, how accurate will its readings be?
A. They will not be accurate
B. They will be accurate enough to get by
C. If it properly calibrates to full scale in the set position, they may be accurate
D. They will be accurate providing the readings are multiplied by 4.5

T4D Dummy antennas, S-meter, exposure of the human body to RF.

T4D01
What device should be connected to a transmitter's output when you are making transmitter adjustments?
A. A multimeter
B. A reflectometer
C. A receiver
D. A dummy antenna

T4D02
What is a dummy antenna?
A. A nondirectional transmitting antenna
B. A nonradiating load for a transmitter
C. An antenna used as a reference for gain measurements
D. A flexible antenna usually used on hand-held transceivers

T4D03
What is the main component of a dummy antenna?
A. A wire-wound resistor
B. An iron-core coil
C. A noninductive resistor
D. An air-core coil

T4D04
What device is used in place of an antenna during transmitter tests so that no signal is radiated?
A. An antenna matcher
B. A dummy antenna
C. A low-pass filter
D. A decoupling resistor

T4D05
Why would you use a dummy antenna?
A. For off-the-air transmitter testing
B. To reduce output power
C. To give comparative signal reports
D. To allow antenna tuning without causing interference

T4D06
What minimum rating should a dummy antenna have for use with a 100-watt single-sideband phone transmitter?
A. 100 watts continuous
B. 141 watts continuous
C. 175 watts continuous
D. 200 watts continuous

T4D07
Why might a dummy antenna get warm when in use?
A. Because it stores electric current
B. Because it stores radio waves
C. Because it absorbs static electricity
D. Because it changes RF energy into heat

T4D08
What is used to measure relative signal strength in a receiver?
A. An S meter
B. An RST meter
C. A signal deviation meter
D. An SSB meter

T4D09
How can exposure to a large amount of RF energy affect body tissue?
A. It causes radiation poisoning
B. It heats the tissue
C. It paralyzes the tissue
D. It produces genetic changes in the tissue

T4D10
Which body organ is the most likely to be damaged from the heating effects of RF radiation?
A. Eyes
B. Hands
C. Heart
D. Liver

T4D11
What organization has published safety guidelines for the maximum limits of RF energy near the human body?
A. The Institute of Electrical and Electronics Engineers (IEEE)
B. The Federal Communications Commission (FCC)
C. The Environmental Protection Agency (EPA)
D. The American National Standards Institute (ANSI)

T4D12
What is the purpose of the ANSI RF protection guide?
A. It lists all RF frequency allocations for interference protection
B. It gives RF exposure limits for the human body
C. It sets transmitter power limits for interference protection
D. It sets antenna height limits for aircraft protection

T4D13
According to the ANSI RF protection guide, what frequencies cause us the greatest risk from RF energy?
A. 3 to 30 MHz
B. 300 to 3000 MHz
C. Above 1500 MHz
D. 30 to 300 MHz

T4D14
Why is the limit of exposure to RF the lowest in the frequency range of 30 MHz to 300 MHz, according to the ANSI RF protection guide?
A. There are more transmitters operating in this range
B. There are fewer transmitters operating in this range
C. Most transmissions in this range are for a longer time
D. The human body absorbs RF energy the most in this range

T4D15
According to the ANSI RF protection guide, what is the maximum safe power output to the antenna of a hand-held VHF or UHF radio?
A. 125 milliwatts
B. 7 watts
C. 10 watts
D. 25 watts

T4D16
After you have opened a VHF power amplifier to make internal tuning adjustments, what should you do before you turn the amplifier on?
A. Remove all amplifier shielding to ensure maximum cooling
B. Make sure that the power interlock switch is bypassed so you can test the amplifier
C. Be certain all amplifier shielding is fastened in place
D. Be certain no antenna is attached so that you will not cause any interference

SUBELEMENT T5—ELECTRICAL PRINCIPLES
[2 exam questions—2 groups]

T5A Definition of resistance, inductance, and capacitance and unit of measurement, calculation of values in series and parallel.

T5A01
What does resistance do in an electric circuit?
A. It stores energy in a magnetic field
B. It stores energy in an electric field
C. It provides electrons by a chemical reaction
D. It opposes the flow of electrons

T5A02
What is the ability to store energy in a magnetic field called?
A. Admittance
B. Capacitance
C. Resistance
D. Inductance

T5A03
What is the basic unit of inductance?
A. The coulomb
B. The farad
C. The henry
D. The ohm

T5A04
What is a henry?
A. The basic unit of admittance
B. The basic unit of capacitance
C. The basic unit of inductance
D. The basic unit of resistance

T5A05
What is the ability to store energy in an electric field called?
A. Inductance
B. Resistance
C. Tolerance
D. Capacitance

T5A06
What is the basic unit of capacitance?
A. The farad
B. The ohm
C. The volt
D. The henry

T5A07
What is a farad?
A. The basic unit of resistance
B. The basic unit of capacitance
C. The basic unit of inductance
D. The basic unit of admittance

T5A08
If two equal-value inductors are connected in series, what is their total inductance?
A. Half the value of one inductor
B. Twice the value of one inductor
C. The same as the value of either inductor
D. The value of one inductor times the value of the other

T5A09
If two equal-value inductors are connected in parallel, what is their total inductance?
A. Half the value of one inductor
B. Twice the value of one inductor
C. The same as the value of either inductor
D. The value of one inductor times the value of the other

T5A10
If two equal-value capacitors are connected in series, what is their total capacitance?
A. Twice the value of one capacitor
B. The same as the value of either capacitor
C. Half the value of either capacitor
D. The value of one capacitor times the value of the other

T5A11
If two equal-value capacitors are connected in parallel, what is their total capacitance?
A. Twice the value of one capacitor
B. Half the value of one capacitor
C. The same as the value of either capacitor
D. The value of one capacitor times the value of the other

T5B Ohm's Law.

T5B01
Ohm's Law describes the mathematical relationship between what three electrical quantities?
A. Resistance, voltage and power
B. Current, resistance and power
C. Current, voltage and power
D. Resistance, current and voltage

T5B02
How is the current in a DC circuit calculated when the voltage and resistance are known?
A. I = R × E [current equals resistance multiplied by voltage]
B. I = R / E [current equals resistance divided by voltage]
C. I = E / R [current equals voltage divided by resistance]
D. I = P / E [current equals power divided by voltage]

T5B03
How is the resistance in a DC circuit calculated when the voltage and current are known?
A. R = I / E [resistance equals current divided by voltage]
B. R = E / I [resistance equals voltage divided by current]
C. R = I × E [resistance equals current multiplied by voltage]
D. R = P / E [resistance equals power divided by voltage]

T5B04
How is the voltage in a DC circuit calculated when the current and resistance are known?
A. E = I / R [voltage equals current divided by resistance]
B. E = R / I [voltage equals resistance divided by current]
C. E = I × R [voltage equals current multiplied by resistance]
D. E = P / I [voltage equals power divided by current]

T5B05
If a 12-volt battery supplies 0.25 ampere to a circuit, what is the circuit's resistance?
A. 0.25 ohm
B. 3 ohm
C. 12 ohms
D. 48 ohms

T5B06
If a 12-volt battery supplies 0.15 ampere to a circuit, what is the circuit's resistance?
A. 0.15 ohm
B. 1.8 ohm
C. 12 ohms
D. 80 ohms

T5B07
If a 4800-ohm resistor is connected to 120 volts, approximately how much current will flow through it?
A. 4 A
B. 25 mA
C. 25 A
D. 40 MA

T5B08
If a 48000-ohm resistor is connected to 120 volts, approximately how much current will flow through it?
A. 400 A
B. 40 A
C. 25 mA
D. 2.5 mA

T5B09
If a 4800-ohm resistor is connected to 12 volts, approximately how much current will flow through it?
A. 2.5 mA
B. 25 mA
C. 40 A
D. 400 A

T5B10
If a 48000-ohm resistor is connected to 12 volts, approximately how much current will flow through it?
A. 250 μA
B. 250 mA
C. 4000 mA
D. 4000 A

T5B11
If you know the voltage and current supplied to a circuit, what formula would you use to calculate the circuit's resistance?
A. Ohm's Law
B. Tesla's Law
C. Ampere's Law
D. Kirchhoff's Law

SUBELEMENT T6—CIRCUIT COMPONENTS
[2 exam questions—2 groups]

T6A Resistors, construction types, variable and fixed, color code, power ratings, schematic symbols.

T6A01
Which of the following are common resistor types?
A. Plastic and porcelain
B. Film and wire-wound
C. Electrolytic and metal-film
D. Iron core and brass core

T6A02
What does a variable resistor or potentiometer do?
A. Its resistance changes when AC is applied to it
B. It transforms a variable voltage into a constant voltage
C. Its resistance changes when its slide or contact is moved
D. Its resistance changes when it is heated

T6A03
How do you find a resistor's tolerance rating?
A. By using a voltmeter
B. By reading the resistor's color code
C. By using Thevenin's theorem for resistors
D. By reading its Baudot code

T6A04
What do the first three color bands on a resistor indicate?
A. The value of the resistor in ohms
B. The resistance tolerance in percent
C. The power rating in watts
D. The resistance material

T6A05
What does the fourth color band on a resistor indicate?
A. The value of the resistor in ohms
B. The resistance tolerance in percent
C. The power rating in watts
D. The resistance material

T6A06
Why do resistors sometimes get hot when in use?
A. Some electrical energy passing through them is lost as heat
B. Their reactance makes them heat up
C. Hotter circuit components nearby heat them up
D. They absorb magnetic energy which makes them hot

T6A07
Why would a large size resistor be used instead of a smaller one of the same resistance?
A. For better response time
B. For a higher current gain
C. For greater power dissipation
D. For less impedance in the circuit

T6A08
What are the possible values of a 100-ohm resistor with a 10% tolerance?
A. 90 to 100 ohms
B. 10 to 100 ohms
C. 90 to 110 ohms
D. 80 to 120 ohms

T6A09
How do you find a resistor's value?
A. By using a voltmeter
B. By using the resistor's color code
C. By using Thevenin's theorem for resistors
D. By using the Baudot code

T6A10
Which tolerance rating would a high-quality resistor have?
A. 0.1%
B. 5%
C. 10%
D. 20%

T6A11
Which tolerance rating would a low-quality resistor have?
A. 0.1%
B. 5%
C. 10%
D. 20%

T6B Schematic symbols—inductors and capacitors, construction of variable and fixed, factors affecting inductance and capacitance, capacitor construction.

T6B01
What is an inductor core?
A. The place where a coil is tapped for resonance
B. A tight coil of wire used in a transformer
C. Insulating material placed between the wires of a transformer
D. The place inside an inductor where its magnetic field is concentrated

T6B02
What does an inductor do?
A. It stores a charge electrostatically and opposes a change in voltage
B. It stores a charge electrochemically and opposes a change in current
C. It stores a charge electromagnetically and opposes a change in current
D. It stores a charge electromechanically and opposes a change in voltage

T6B03
What determines the inductance of a coil?
A. The core material, the core diameter, the length of the coil and whether the coil is mounted horizontally or vertically
B. The core diameter, the number of turns of wire used to wind the coil and the type of metal used for the wire
C. The core material, the number of turns used to wind the core and the frequency of the current through the coil
D. The core material, the core diameter, the length of the coil and the number of turns of wire used to wind the coil

T6B04
As an iron core is inserted in a coil, what happens to the coil's inductance?
A. It increases
B. It decreases
C. It stays the same
D. It disappears

T6B05
What can happen if you tune a ferrite-core coil with a metal tool?
A. The metal tool can change the coil's inductance and cause you to tune the coil incorrectly
B. The metal tool can become magnetized so much that you might not be able to remove it from the coil
C. The metal tool can pick up enough magnetic energy to become very hot
D. The metal tool can pick up enough magnetic energy to become a shock hazard

T6B06
In Figure T6-1 which symbol represents an adjustable inductor?
A. Symbol 1
B. Symbol 2
C. Symbol 3
D. Symbol 4

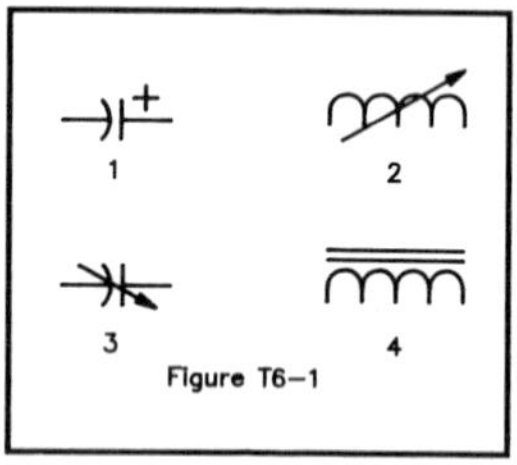

Figure T6–1

T6B07
In Figure T6-1 which symbol represents an iron-core inductor?
A. Symbol 1
B. Symbol 2
C. Symbol 3
D. Symbol 4

T6B08
In Figure T6-1 which symbol represents an inductor wound over a toroidal core?
A. Symbol 1
B. Symbol 2
C. Symbol 3
D. Symbol 4

T6B09
In Figure T6-1 which symbol represents an electrolytic capacitor?
A. Symbol 1
B. Symbol 2
C. Symbol 3
D. Symbol 4

T6B10
In Figure T6-1 which symbol represents a variable capacitor?
A. Symbol 1
B. Symbol 2
C. Symbol 3
D. Symbol 4

T6B11
What describes a capacitor?
A. Two or more layers of silicon material with an insulating material between them
B. The wire used in the winding and the core material
C. Two or more conductive plates with an insulating material between them
D. Two or more insulating plates with a conductive material between them

T6B12
What does a capacitor do?
A. It stores a charge electrochemically and opposes a change in current
B. It stores a charge electrostatically and opposes a change in voltage
C. It stores a charge electromagnetically and opposes a change in current
D. It stores a charge electromechanically and opposes a change in voltage

T6B13
What determines the capacitance of a capacitor?
A. The material between the plates, the area of one side of one plate, the number of plates and the spacing between the plates
B. The material between the plates, the number of plates and the size of the wires connected to the plates
C. The number of plates, the spacing between the plates and whether the dielectric material is N type or P type
D. The material between the plates, the area of one plate, the number of plates and the material used for the protective coating

T6B14
As the plate area of a capacitor is increased, what happens to its capacitance?
A. It decreases
B. It increases
C. It stays the same
D. It disappears

SUBELEMENT T7—PRACTICAL CIRCUITS
[1 exam question—1 group]

T7A Practical circuits.

T7A01
Why do modern HF transmitters have a built-in low-pass filter in their RF output circuits?
A. To reduce RF energy below a cutoff point
B. To reduce low-frequency interference to other amateurs
C. To reduce harmonic radiation
D. To reduce fundamental radiation

T7A02
What circuit blocks RF energy above and below a certain limit?
A. A band-pass filter
B. A high-pass filter
C. An input filter
D. A low-pass filter

T7A03
What type of filter is used in the IF section of receivers to block energy outside a certain frequency range?
A. A band-pass filter
B. A high-pass filter
C. An input filter
D. A low-pass filter

T7A04
What circuit is found in all types of receivers?
A. An audio filter
B. A beat-frequency oscillator
C. A detector
D. An RF amplifier

T7A05
What circuit has a variable-frequency oscillator connected to a driver and a power amplifier?
A. A packet-radio transmitter
B. A crystal-controlled transmitter
C. A single-sideband transmitter
D. A VFO-controlled transmitter

T7A06
What circuit combines signals from an IF amplifier stage and a beat-frequency oscillator (BFO), to produce an audio signal?
A. An AGC circuit
B. A detector circuit
C. A power supply circuit
D. A VFO circuit

T7A07
What circuit uses a limiter and a frequency discriminator to produce an audio signal?
A. A double-conversion receiver
B. A variable-frequency oscillator
C. A superheterodyne receiver
D. An FM receiver

T7A08
What circuit is pictured in Figure T7-1 if block 1 is a variable-frequency oscillator?
A. A packet-radio transmitter
B. A crystal-controlled transmitter
C. A single-sideband transmitter
D. A VFO-controlled transmitter

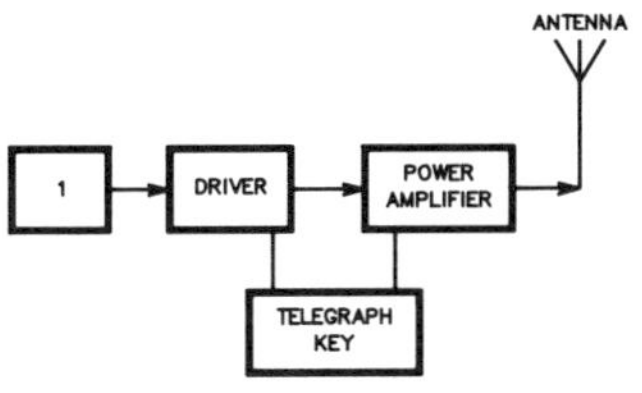

Figure T7–1

T7A09
What is the unlabeled block in Figure T7-2?
A. An AGC circuit
B. A detector
C. A power supply
D. A VFO circuit

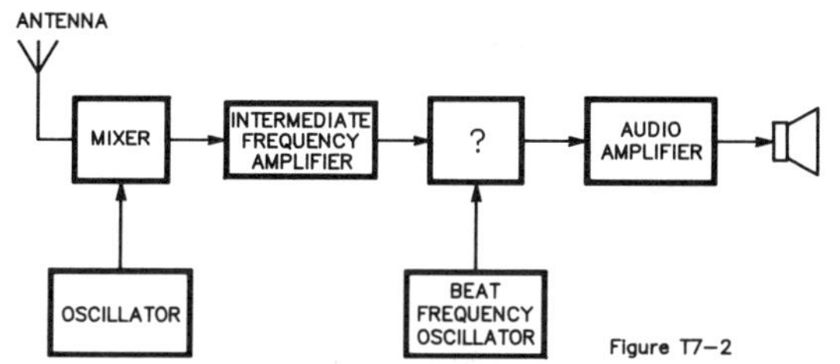

Figure T7–2

T7A10
What circuit is pictured in Figure T7-3?
A. A double-conversion receiver
B. A variable-frequency oscillator
C. A superheterodyne receiver
D. An FM receiver

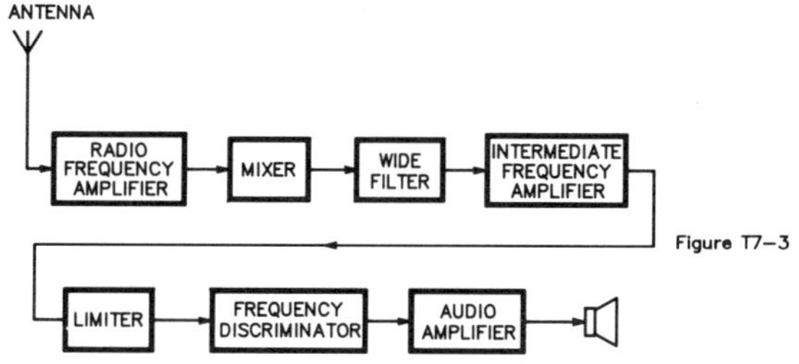

Figure T7–3

T7A11
What is the unlabeled block in Figure T7-4?
A. A band-pass filter
B. A crystal oscillator
C. A reactance modulator
D. A rectifier modulator

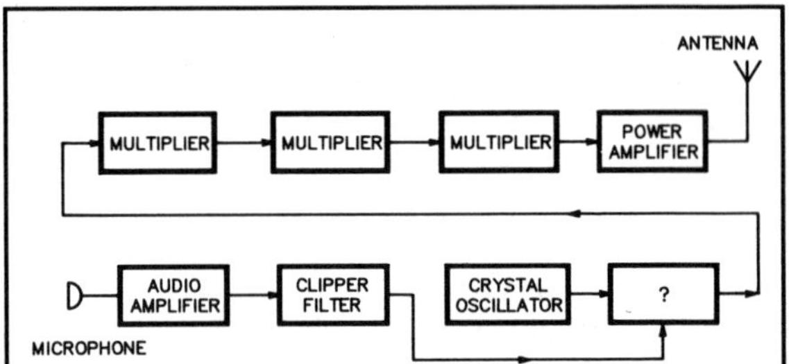

Figure T7-4

SUBELEMENT T8—SIGNALS AND EMISSIONS
[2 exam questions—2 groups]

T8A Definition of modulation and emission types.

T8A01
What is the name for unmodulated carrier wave emissions?
A. Phone
B. Test
C. CW
D. RTTY

T8A02
What is the name for Morse code emissions produced by switching a transmitter's output on and off?
A. Phone
B. Test
C. CW
D. RTTY

T8A03
What is RTTY?
A. Amplitude-keyed telegraphy
B. Frequency-shift-keyed telegraphy
C. Frequency-modulated telephony
D. Phase-modulated telephony

T8A04
What is the name for packet-radio emissions?
A. CW
B. Data
C. Phone
D. RTTY

T8A05

How is tone-modulated Morse code produced?

A. By feeding a microphone's audio signal into an FM transmitter
B. By feeding an on/off keyed audio tone into a CW transmitter
C. By on/off keying of a carrier
D. By feeding an on/off keyed audio tone into a transmitter

T8A06

What is the name of the voice emission most used on VHF/UHF repeaters?

A. Single-sideband phone
B. Pulse-modulated phone
C. Slow-scan phone
D. Frequency-modulated phone

T8A07

What is the name of the voice emission most used on amateur HF bands?

A. Single-sideband phone
B. Pulse-modulated phone
C. Slow-scan phone
D. Frequency-modulated phone

T8A08

What is meant by the upper sideband (USB)?

A. The part of a single-sideband signal which is above the carrier frequency
B. The part of a single-sideband signal which is below the carrier frequency
C. Any frequency above 10 MHz
D. The carrier frequency of a single-sideband signal

T8A09

What emissions are produced by a transmitter using a reactance modulator?

A. CW
B. Test
C. Single-sideband, suppressed-carrier phone
D. Phase-modulated phone

T8A10

What other emission does phase modulation most resemble?

A. Amplitude modulation
B. Pulse modulation
C. Frequency modulation
D. Single-sideband modulation

T8A11

What is the name for emissions produced by an on/off keyed audio tone?

A. RTTY
B. MCW
C. CW
D. Phone

T8B RF carrier, modulation, bandwidth and deviation.

T8B01
What is another name for a constant-amplitude radio-frequency signal?
A. An RF carrier
B. An AF carrier
C. A sideband carrier
D. A subcarrier

T8B02
What is modulation?
A. Varying a radio wave in some way to send information
B. Receiving audio information from a signal
C. Increasing the power of a transmitter
D. Suppressing the carrier in a single-sideband transmitter

T8B03
What kind of emission would your FM transmitter produce if its microphone failed to work?
A. An unmodulated carrier
B. A phase-modulated carrier
C. An amplitude-modulated carrier
D. A frequency-modulated carrier

T8B04
How would you modulate a 2-meter FM transceiver to produce packet-radio emissions?
A. Connect a terminal-node controller to interrupt the transceiver's carrier wave
B. Connect a terminal-node controller to the transceiver's microphone input
C. Connect a keyboard to the transceiver's microphone input
D. Connect a DTMF key pad to the transceiver's microphone input

T8B05
Why is FM voice best for local VHF/UHF radio communications?
A. The carrier is not detectable
B. It is more resistant to distortion caused by reflected signals
C. It has high-fidelity audio which can be understood even when the signal is somewhat weak
D. Its RF carrier stays on frequency better than the AM modes

T8B06
Why do many radio receivers have several IF filters of different bandwidths that can be selected by the operator?
A. Because some frequency bands are wider than others
B. Because different bandwidths help increase the receiver sensitivity
C. Because different bandwidths improve S-meter readings
D. Because some emission types need a wider bandwidth than others to be received properly

T8B07
Which list of emission types is in order from the narrowest bandwidth to the widest bandwidth?
A. RTTY, CW, SSB voice, FM voice
B. CW, FM voice, RTTY, SSB voice
C. CW, RTTY, SSB voice, FM voice
D. CW, SSB voice, RTTY, FM voice

T8B08
What is the usual bandwidth of a single-sideband amateur signal?
A. 1 kHz
B. 2 kHz
C. Between 3 and 6 kHz
D. Between 2 and 3 kHz

T8B09
What is the usual bandwidth of a frequency-modulated amateur signal?
A. Less than 5 kHz
B. Between 5 and 10 kHz
C. Between 10 and 20 kHz
D. Greater than 20 kHz

T8B10
What is the result of overdeviation in an FM transmitter?
A. Increased transmitter power
B. Out-of-channel emissions
C. Increased transmitter range
D. Poor carrier suppression

T8B11
What causes splatter interference?
A. Keying a transmitter too fast
B. Signals from a transmitter's output circuit are being sent back to its input circuit
C. Overmodulation of a transmitter
D. The transmitting antenna is the wrong length

SUBELEMENT T9—ANTENNAS AND FEED LINES
[3 exam questions—3 groups]

T9A Parasitic beam and non-directional antennas.

T9A01
What is a directional antenna?
A. An antenna which sends and receives radio energy equally well in all directions
B. An antenna that cannot send and receive radio energy by skywave or skip propagation
C. An antenna which sends and receives radio energy mainly in one direction
D. An antenna which sends and receives radio energy equally well in two opposite directions

T9A02
How is a Yagi antenna constructed?
A. Two or more straight, parallel elements are fixed in line with each other
B. Two or more square or circular loops are fixed in line with each other
C. Two or more square or circular loops are stacked inside each other
D. A straight element is fixed in the center of three or more elements which angle toward the ground

T9A03
What type of beam antenna uses two or more straight elements arranged in line with each other?
A. A delta loop antenna
B. A quad antenna
C. A Yagi antenna
D. A Zepp antenna

T9A04
How many directly driven elements do most beam antennas have?
A. None
B. One
C. Two
D. Three

T9A05
What is a parasitic beam antenna?
A. An antenna where some elements obtain their radio energy by induction or radiation from a driven element
B. An antenna where wave traps are used to magnetically couple the elements
C. An antenna where all elements are driven by direct connection to the feed line
D. An antenna where the driven element obtains its radio energy by induction or radiation from director elements

T9A06
What are the parasitic elements of a Yagi antenna?
A. The driven element and any reflectors
B. The director and the driven element
C. Only the reflectors (if any)
D. Any directors or any reflectors

T9A07
What is a cubical quad antenna?
A. Four straight, parallel elements in line with each other, each approximately 1/2-electrical wavelength long
B. Two or more parallel four-sided wire loops, each approximately one-electrical wavelength long
C. A vertical conductor 1/4-electrical wavelength high, fed at the bottom
D. A center-fed wire 1/2-electrical wavelength long

T9A08
What is a delta loop antenna?
A. A type of cubical quad antenna, except with triangular elements rather than square
B. A large copper ring or wire loop, used in direction finding
C. An antenna system made of three vertical antennas, arranged in a triangular shape
D. An antenna made from several triangular coils of wire on an insulating form

T9A09
What type of non-directional antenna is easy to make at home and works well outdoors?
A. A Yagi
B. A delta loop
C. A cubical quad
D. A ground plane

T9A10
What type of antenna is made when a magnetic-base whip antenna is placed on the roof of a car?
A. A Yagi
B. A delta loop
C. A cubical quad
D. A ground plane

T9A11
If a magnetic-base whip antenna is placed on the roof of a car, in what direction does it send out radio energy?
A. It goes out equally well in all horizontal directions
B. Most of it goes in one direction
C. Most of it goes equally in two opposite directions
D. Most of it is aimed high into the air

T9B Polarization, impedance matching and SWR, feed lines, balanced vs unbalanced (including baluns).

T9B01
What does horizontal wave polarization mean?
A. The magnetic lines of force of a radio wave are parallel to the earth's surface
B. The electric lines of force of a radio wave are parallel to the earth's surface
C. The electric lines of force of a radio wave are perpendicular to the earth's surface
D. The electric and magnetic lines of force of a radio wave are perpendicular to the earth's surface

T9B02
What does vertical wave polarization mean?
A. The electric lines of force of a radio wave are parallel to the earth's surface
B. The magnetic lines of force of a radio wave are perpendicular to the earth's surface
C. The electric lines of force of a radio wave are perpendicular to the earth's surface
D. The electric and magnetic lines of force of a radio wave are parallel to the earth's surface

T9B03
What electromagnetic-wave polarization does a Yagi antenna have when its elements are parallel to the earth's surface?
A. Circular
B. Helical
C. Horizontal
D. Vertical

T9B04
What electromagnetic-wave polarization does a half-wavelength antenna have when it is perpendicular to the earth's surface?
A. Circular
B. Horizontal
C. Parabolical
D. Vertical

T9B05
What electromagnetic-wave polarization does most man-made electrical noise have in the HF and VHF spectrum?
A. Horizontal
B. Left-hand circular
C. Right-hand circular
D. Vertical

T9B06
What does standing-wave ratio mean?
A. The ratio of maximum to minimum inductances on a feed line
B. The ratio of maximum to minimum resistances on a feed line
C. The ratio of maximum to minimum impedances on a feed line
D. The ratio of maximum to minimum voltages on a feed line

T9B07
What does forward power mean?
A. The power traveling from the transmitter to the antenna
B. The power radiated from the top of an antenna system
C. The power produced during the positive half of an RF cycle
D. The power used to drive a linear amplifier

T9B08
What does reflected power mean?
A. The power radiated down to the ground from an antenna
B. The power returned to a transmitter from an antenna
C. The power produced during the negative half of an RF cycle
D. The power returned to an antenna by buildings and trees

T9B09
What happens to radio energy when it is sent through a poor quality coaxial cable?
A. It causes spurious emissions
B. It is returned to the transmitter's chassis ground
C. It is converted to heat in the cable
D. It causes interference to other stations near the transmitting frequency

T9B10
What is an unbalanced line?
A. Feed line with neither conductor connected to ground
B. Feed line with both conductors connected to ground
C. Feed line with one conductor connected to ground
D. Feed line with both conductors connected to each other

T9B11
What device can be installed to feed a balanced antenna with an unbalanced feed line?
A. A balun
B. A loading coil
C. A triaxial transformer
D. A wavetrap

T9C Line losses by line type, length and frequency, RF safety.

T9C01
What common connector usually joins RG-213 coaxial cable to an HF transceiver?
A. An F-type cable connector
B. A PL-259 connector
C. A banana plug connector
D. A binding post connector

T9C02
What common connector usually joins a hand-held transceiver to its antenna?
A. A BNC connector
B. A PL-259 connector
C. An F-type cable connector
D. A binding post connector

T9C03
Which of these common connectors has the lowest loss at UHF?
A. An F-type cable connector
B. A type-N connector
C. A BNC connector
D. A PL-259 connector

T9C04
If you install a 6-meter Yagi antenna on a tower 150 feet from your transmitter, which of the following feed lines is best?
A. RG-213
B. RG-58
C. RG-59
D. RG-174

T9C05
If you have a transmitter and an antenna which are 50 feet apart, but are connected by 200 feet of RG-58 coaxial cable, what should be done to reduce feed line loss?
A. Cut off the excess cable so the feed line is an even number of wavelengths long
B. Cut off the excess cable so the feed line is an odd number of wavelengths long
C. Cut off the excess cable
D. Roll the excess cable into a coil which is as small as possible

T9C06
As the length of a feed line is changed, what happens to signal loss?
A. Signal loss is the same for any length of feed line
B. Signal loss increases as length increases
C. Signal loss decreases as length increases
D. Signal loss is the least when the length is the same as the signal's wavelength

T9C07
As the frequency of a signal is changed, what happens to signal loss in a feed line?
A. Signal loss is the same for any frequency
B. Signal loss increases with increasing frequency
C. Signal loss increases with decreasing frequency
D. Signal loss is the least when the signal's wavelength is the same as the feed line's length

T9C08
If your antenna feed line gets hot when you are transmitting, what might this mean?
A. You should transmit using less power
B. The conductors in the feed line are not insulated very well
C. The feed line is too long
D. The SWR may be too high, or the feed line loss may be high

T9C09
Why should you make sure that no one can touch an open-wire feed line while you are transmitting with it?
A. Because contact might cause a short circuit and damage the transmitter
B. Because contact might break the feed line
C. Because contact might cause spurious emissions
D. Because high-voltage radio energy might burn the person

T9C10
For RF safety, what is the best thing to do with your transmitting antennas?
A. Use vertical polarization
B. Use horizontal polarization
C. Mount the antennas where no one can come near them
D. Mount the antenna close to the ground

T9C11
Why should you regularly clean, tighten and re-solder all antenna connectors?
A. To help keep their resistance at a minimum
B. To keep them looking nice
C. To keep them from getting stuck in place
D. To increase their capacitance

ANSWER KEY
Page numbers tell you where to look in this book for more information.

Subelement T1

Numbers in [square brackets] indicate sections in FCC Part 97, the Amateur Radio rules.

T1A01 (D) [97.3a12]
T1A02 (B) [97.3a12]
T1A03 (A) [97.19a/b]
T1A04 (A) [97.19c]
T1A05 (C) [97.301/305e]
T1A06 (C) [97.301a]
T1A07 (B) [97.301a]
T1A08 (A) [97.301a]
T1A09 (B) [97.301a]
T1A10 (D) [97.301e]
T1A11 (C) [97.301e]
T1B01 (C) [97.3b6]
T1B02 (D) [97.3b6]
T1B03 (B) [97.203c]
T1B04 (C) [97.303]
T1B05 (D) [97.303]
T1B06 (C) [97.101(b)]
T1B07 (A) [97.305a]
T1B08 (B) [97.305c]
T1B09 (A) [97.305c]
T1B10 (D) [97.313b]
T1B11 (A) [97.209b2]
T1C01 (A) [97.119e1]
T1C02 (C) [97.307f3/4]
T1C03 (D) [97.307]
T1C04 (B) [97.307f4]
T1C05 (C) [97.307f5]
T1C06 (C) [97.307f4]
T1C07 (B) [97.307f5]
T1C08 (A) [97.307f5]
T1C09 (D) [97.307f6]
T1C10 (B) [97.307f6]
T1C11 (C) [97.307f6]
T1D01 (A) [97.3a9]
T1D02 (B) [97.119b1]
T1D03 (C) [97.119b2]
T1D04 (C) [97.119b2]
T1D05 (B) [97.203a]
T1D06 (A) [97.205c]
T1D07 (D) [97.205c]
T1D08 (A) [97.205c]
T1D09 (C) [97.215a]
T1D10 (D) [97.215a]
T1D11 (B) [97.215c]
T1E01 (A) [97.3a10]
T1E02 (B) [97.3a10]
T1E03 (D) [97.113b]
T1E04 (B) [97.113d]
T1E05 (D) [97.113d]
T1E06 (C) [97.113d]
T1E07 (C) [97.113e]
T1E08 (D) [97.115a2]
T1E09 (A) [97.115b1]
T1E10 (A) [97.401a]
T1E11 (C) [97.401c]

Subelement T2
T2A01 (B) p 2-6
T2A02 (C) p 2-6
T2A03 (A) p 2-6
T2A04 (D) p 2-6
T2A05 (B) p 2-4
T2A06 (B) p 2-7
T2A07 (D) p 2-7
T2A08 (B) p 2-6
T2A09 (A) p 2-4
T2A10 (C) p 2-4
T2A11 (A) p 2-4
T2A12 (A) p 2-5
T2A13 (C) p 2-5
T2A14 (D) p 2-5
T2A15 (A) p 2-4
T2A16 (A) p 2-4
T2A17 (B) p 2-1
T2A18 (C) p 2-6
T2B01 (C) p 2-7
T2B02 (A) p 2-7
T2B03 (C) p 2-7
T2B04 (D) p 2-5
T2B05 (C) p 2-1
T2B06 (A) p 2-2
T2B07 (B) p 2-2
T2B08 (B) p 2-2
T2B09 (A) p 2-3
T2B10 (C) p 2-3
T2B11 (D) p 2-3
T2C01 (A) p 2-8
T2C02 (D) p 2-8
T2C03 (A) p 2-8
T2C04 (C) p 2-8
T2C05 (A) p 2-8
T2C06 (D) p 2-8
T2C07 (B) p 2-9
T2C08 (D) p 2-9
T2C09 (B) p 2-9
T2C10 (B) p 2-9
T2C11 (C) p 2-9
T2C12 (C) p 2-9

Subelement T3
T3A01 (A) p 3-1
T3A02 (D) p 3-1
T3A03 (A) p 3-1
T3A04 (C) p 3-1
T3A05 (A) p 3-2
T3A06 (B) p 3-2
T3A07 (B) p 3-4
T3A08 (A) p 3-2
T3A09 (D) p 3-4
T3A10 (B) p 3-4
T3A11 (C) p 3-4
T3B01 (D) p 3-2
T3B02 (B) p 3-2
T3B03 (A) p 3-2
T3B04 (B) p 3-2
T3B05 (D) p 3-4
T3B06 (C) p 3-4
T3B07 (A) p 3-4
T3B08 (B) p 3-4
T3B09 (A) p 3-4
T3B10 (C) p 3-6
T3B11 (A) p 3-6
T3C01 (C) p 3-7
T3C02 (C) p 3-9
T3C03 (B) p 3-9
T3C04 (C) p 3-10
T3C05 (A) p 3-10
T3C06 (D) p 3-10
T3C07 (B) p 3-10
T3C08 (D) p 3-10
T3C09 (D) p 3-10
T3C10 (A) p 3-10
T3C11 (D) p 3-7

Subelement T4
T4A01 (C) p 4-2
T4A02 (D) p 4-2
T4A03 (B) p 4-2

T4A04 (D) p 4-4
T4A05 (C) p 4-4
T4A06 (D) p 4-5
T4A07 (B) p 4-3
T4A08 (A) p 4-7
T4A09 (A) p 4-8
T4A10 (A) p 4-8
T4A11 (A) p 4-7
T4A12 (C) p 4-6
T4A13 (B) p 4-5
T4A14 (C) p 4-8
T4A15 (B) p 4-8
T4B01 (B) p 4-8
T4B02 (C) p 4-9
T4B03 (A) p 4-9
T4B04 (A) p 4-9
T4B05 (D) p 4-9
T4B06 (D) p 4-10
T4B07 (A) p 4-11
T4B08 (B) p 4-11
T4B09 (A) p 4-12
T4B10 (B) p 4-12
T4B11 (C) p 4-12
T4C01 (A) p 4-12
T4C02 (A) p 4-12
T4C03 (D) p 4-12
T4C04 (B) p 4-13
T4C05 (C) p 4-13
T4C06 (C) p 4-13
T4C07 (B) p 4-13
T4C08 (D) p 4-14
T4C09 (A) p 4-14
T4C10 (A) p 4-11
T4C11 (C) p 4-14
T4D01 (D) p 4-14
T4D02 (B) p 4-14
T4D03 (C) p 4-15
T4D04 (B) p 4-15
T4D05 (A) p 4-15
T4D06 (A) p 4-15
T4D07 (D) p 4-15
T4D08 (A) p 4-15
T4D09 (B) p 4-16
T4D10 (A) p 4-16
T4D11 (D) p 4-16
T4D12 (B) p 4-16
T4D13 (D) p 4-17
T4D14 (D) p 4-17
T4D15 (B) p 4-16
T4D16 (C) p 4-18

Subelement T5

T5A01 (D) p 5-4
T5A02 (D) p 5-9
T5A03 (C) p 5-10
T5A04 (C) p 5-10
T5A05 (D) p 5-12
T5A06 (A) p 5-12
T5A07 (B) p 5-12
T5A08 (B) p 5-10
T5A09 (A) p 5-11
T5A10 (C) p 5-14
T5A11 (A) p 5-13
T5B01 (D) p 5-6
T5B02 (C) p 5-6
T5B03 (B) p 5-6
T5B04 (C) p 5-6
T5B05 (D) p 5-7
T5B06 (D) p 5-7
T5B07 (B) p 5-7
T5B08 (D) p 5-7
T5B09 (A) p 5-7
T5B10 (A) p 5-10
T5B11 (A) p 5-6

Subelement T6

T6A01 (B) p 6-1
T6A02 (C) p 6-4
T6A03 (B) p 6-4
T6A04 (A) p 6-4
T6A05 (B) p 6-4
T6A06 (A) p 6-5
T6A07 (C) p 6-5
T6A08 (C) p 6-5
T6A09 (B) p 6-4
T6A10 (A) p 6-5
T6A11 (D) p 6-5
T6B01 (D) p 6-6
T6B02 (C) p 6-6
T6B03 (D) p 6-6
T6B04 (A) p 6-6
T6B05 (A) p 6-7
T6B06 (B) p 6-7
T6B07 (D) p 6-7
T6B08 (D) p 6-8
T6B09 (A) p 6-11
T6B10 (C) p 6-11
T6B11 (C) p 6-9
T6B12 (B) p 6-9
T6B13 (A) p 6-10
T6B14 (B) p 6-10

Subelement T7

T7A01 (C) p 7-3
T7A02 (A) p 7-5
T7A03 (A) p 7-5
T7A04 (C) p 7-10
T7A05 (D) p 7-6
T7A06 (B) p 7-12
T7A07 (D) p 7-12
T7A08 (D) p 7-6
T7A09 (B) p 7-12
T7A10 (D) p 7-12
T7A11 (C) p 7-7

Subelement T8

T8A01 (B) p 8-2
T8A02 (C) p 8-2
T8A03 (B) p 8-4
T8A04 (B) p 8-4
T8A05 (D) p 8-2
T8A06 (D) p 2-4
T8A07 (A) p 8-2
T8A08 (A) p 8-2
T8A09 (D) p 7-9
T8A10 (C) p 7-7
T8A11 (B) p 8-2
T8B01 (A) p 7-7
T8B02 (A) p 7-7
T8B03 (A) p 7-8
T8B04 (B) p 8-4
T8B05 (C) p 8-3
T8B06 (D) p 8-5
T8B07 (C) p 8-5
T8B08 (D) p 8-5
T8B09 (C) p 8-6
T8B10 (B) p 8-7
T8B11 (C) p 8-7

Subelement T9

T9A01 (C) p 9-9
T9A02 (A) p 9-11
T9A03 (C) p 9-11
T9A04 (B) p 9-9
T9A05 (A) p 9-9
T9A06 (D) p 9-11
T9A07 (B) p 9-11
T9A08 (A) p 9-13
T9A09 (D) p 9-9
T9A10 (D) p 9-9
T9A11 (A) p 9-9
T9B01 (B) p 9-14
T9B02 (C) p 9-14
T9B03 (C) p 9-15
T9B04 (D) p 9-15
T9B05 (D) p 9-15
T9B06 (D) p 9-8
T9B07 (A) p 9-8
T9B08 (B) p 9-8
T9B09 (C) p 9-3
T9B10 (C) p 9-7
T9B11 (A) p 9-7
T9C01 (B) p 9-5
T9C02 (A) p 9-5
T9C03 (B) p 9-6
T9C04 (A) p 9-5
T9C05 (C) p 9-5
T9C06 (B) p 9-5
T9C07 (B) p 9-5
T9C08 (D) p 9-3
T9C09 (D) p 9-16
T9C10 (C) p 9-15
T9C11 (A) p 9-6

A Useful Tables

US Customary—Metric Conversion Factors

International System of Units (SI)—Metric Units

Prefix	Symbol			Multiplication Factor
exa	E	10^{18}	=	1 000 000 000 000 000 000
peta	P	10^{15}	=	1 000 000 000 000 000
tera	T	10^{12}	=	1 000 000 000 000
giga	G	10^{9}	=	1 000 000 000
mega	M	10^{6}	=	1 000 000
kilo	k	10^{3}	=	1 000
hecto	h	10^{2}	=	100
deca	da	10^{1}	=	10
(unit)		10^{0}	=	1
deci	d	10^{-1}	=	0.1
centi	c	10^{-2}	=	0.01
milli	m	10^{-3}	=	0.001
micro	μ	10^{-6}	=	0.000001
nano	n	10^{-9}	=	0.000000001
pico	p	10^{-12}	=	0.000000000001
femto	f	10^{-15}	=	0.000000000000001
atto	a	10^{-18}	=	0.000000000000000001

Linear

1 meter (m) = 100 centimeters (cm) = 1000 millimeters (mm)

Area

1 m^2 = 1 × 10^4 cm^2 = 1 × 10^6 mm^2

Volume

1 m^3 = 1 × 10^6 cm^3 = 1 × 10^9 mm^3
1 litre (l) = 1000 cm^3 = 1 × 10^6 mm^3

Mass

1 kilogram (kg) = 1 000 grams (g)
(Approximately the mass of 1 liter of water)
1 metric ton (or tonne) = 1 000 kg

US Customary Units

Linear Units

12 inches (in) = 1 foot (ft)
36 inches = 3 feet = 1 yard (yd)
1 rod = 5½ yards = 16½ feet
1 statute mile = 1 760 yards = 5 280 feet
1 nautical mile = 6 076.11549 feet

Area

1 ft^2 = 144 in^2
1 yd^2 = 9 ft^2 = 1 296 in^2
1 rod^2 = 30¼ yd^2
1 acre = 4840 yd^2 = 43 560 ft^2
1 acre = 160 rod^2
1 $mile^2$ = 640 acres

Volume

1 ft^3 = 1 728 in^3
1 yd^3 = 27 ft^3

Liquid Volume Measure

1 fluid ounce (fl oz) = 8 fluidrams = 1.804 in^3
1 pint (pt) = 16 fl oz
1 quart (qt) = 2 pt = 32 fl oz = 57¾ in^3
1 gallon (gal) = 4 qt = 231 in^3
1 barrel = 31½ gal

Dry Volume Measure

1 quart (qt) = 2 pints (pt) = 67.2 in^3
1 peck = 8 qt
1 bushel = 4 pecks = 2 150.42 in^3

Avoirdupois Weight

1 dram (dr) = 27.343 grains (gr) or (gr a)
1 ounce (oz) = 437.5 gr
1 pound (lb) = 16 oz = 7 000 gr
1 short ton = 2 000 lb, 1 long ton = 2 240 lb

Troy Weight

1 grain troy (gr t) = 1 grain avoirdupois
1 pennyweight (dwt) or (pwt) = 24 gr t
1 ounce troy (oz t) = 480 grains
1 lb t = 12 oz t = 5 760 grains

Apothecaries' Weight

1 grain apothecaries' (gr ap) = 1 gr t = 1 gr a
1 dram ap (dr ap) = 60 gr
1 oz ap = 1 oz t = 8 dr ap = 480 fr
1 lb ap = 1 lb t = 12 oz ap = 5 760 gr

Temperature

°F = 9/5°C + 32
°C = 5/9 (°F − 32)
K = °C + 273
°C = K − 273

Multiply ⟶

Metric Unit = Conversion Factor × US Customary Unit

⟵ Divide

Metric Unit ÷ Conversion Factor = US Customary Unit

Metric Unit =	Conversion Factor ×	US Unit
(Length)		
mm	25.4	inch
cm	2.54	inch
cm	30.48	foot
m	0.3048	foot
m	0.9144	yard
km	1.609	mile
km	1.852	nautical mile
(Area)		
mm^2	645.16	$inch^2$
cm^2	6.4516	in^2
cm^2	929.03	ft^2
m^2	0.0929	ft^2
cm^2	8361.3	yd^2
m^2	0.83613	yd^2
m^2	4047	acre
km^2	2.59	mi^2
(Mass)	(Avoirdupois Weight)	
grams	0.0648	grains
g	28.349	oz
g	453.59	lb
kg	0.45359	lb
tonne	0.907	short ton
tonne	1.016	long ton

Metric Unit =	Conversion Factor ×	US Unit
(Volume)		
mm^3	16387.064	in^3
cm^3	16.387	in^3
m^3	0.028316	ft^3
m^3	0.764555	yd^3
ml	16.387	in^3
ml	29.57	fl oz
ml	473	pint
ml	946.333	quart
l	28.32	ft^3
l	0.9463	quart
l	3.785	gallon
l	1.101	dry quart
l	8.809	peck
l	35.238	bushel
(Mass)	(Troy Weight)	
g	31.103	oz t
g	373.248	lb t
(Mass)	(Apothecaries' Weight)	
g	3.387	dr ap
g	31.103	oz ap
g	373.248	lb ap

Standard Resistance Values

Numbers in **bold** type are ± 10% values. Others are 5% values.

Ohms										*Megohms*				
1.0	3.6	**12**	43	**150**	510	**1800**	6200	**22000**	75000	0.24	0.62	1.6	4.3	11.0
1.1	**3.9**	13	**47**	160	**560**	2000	**6800**	24000	**82000**	**0.27**	**0.68**	**1.8**	**4.7**	**12.0**
1.2	4.3	**15**	51	**180**	620	**2200**	7500	**27000**	91000	0.30	0.75	2.0	5.1	13.0
1.3	**4.7**	16	**56**	200	**680**	2400	**8200**	30000	**100000**	**0.33**	**0.82**	**2.2**	**5.6**	**15.0**
1.5	5.1	**18**	62	**220**	750	**2700**	9100	**33000**	110000	0.36	0.91	2.4	6.2	16.0
1.6	**5.6**	20	**68**	240	**820**	3000	**10000**	36000	**120000**	**0.39**	**1.0**	**2.7**	**6.8**	**18.0**
1.8	6.2	**22**	75	**270**	910	**3300**	11000	**39000**	130000	0.43	1.1	3.0	7.5	20.0
2.0	**6.8**	24	**82**	300	**1000**	3600	**12000**	43000	**150000**	**0.47**	**1.2**	**3.3**	**8.2**	**22.0**
2.2	7.5	**27**	91	**330**	1100	**3900**	13000	**47000**	160000	0.51	1.3	3.6	9.1	
2.4	**8.2**	30	**100**	360	**1200**	4300	**15000**	51000	**180000**	**0.56**	**1.5**	**3.9**	**10.0**	
2.7	9.1	**33**	110	**390**	1300	**4700**	16000	**56000**	200000					
3.0	**10.0**	36	**120**	430	**1500**	5100	**18000**	62000	**220000**					
3.3	11.0	**39**	130	**470**	1600	**5600**	20000	**68000**						

Resistor Color Code

Color	*Sig. Figure*	*Decimal Multiplier*	*Tolerance (%)*
Black	0	1	
Brown	1	10	
Red	2	100	
Orange	3	1,000	
Yellow	4	10,000	
Green	5	100,000	
Blue	6	1,000,000	
Violet	7	10,000,000	
Gray	8	100,000,000	
White	9	1,000,000,000	
Gold	—	0.1	5
Silver	—	0.01	10
No color	—		20

Schematic Symbols Used in Circuit Diagrams

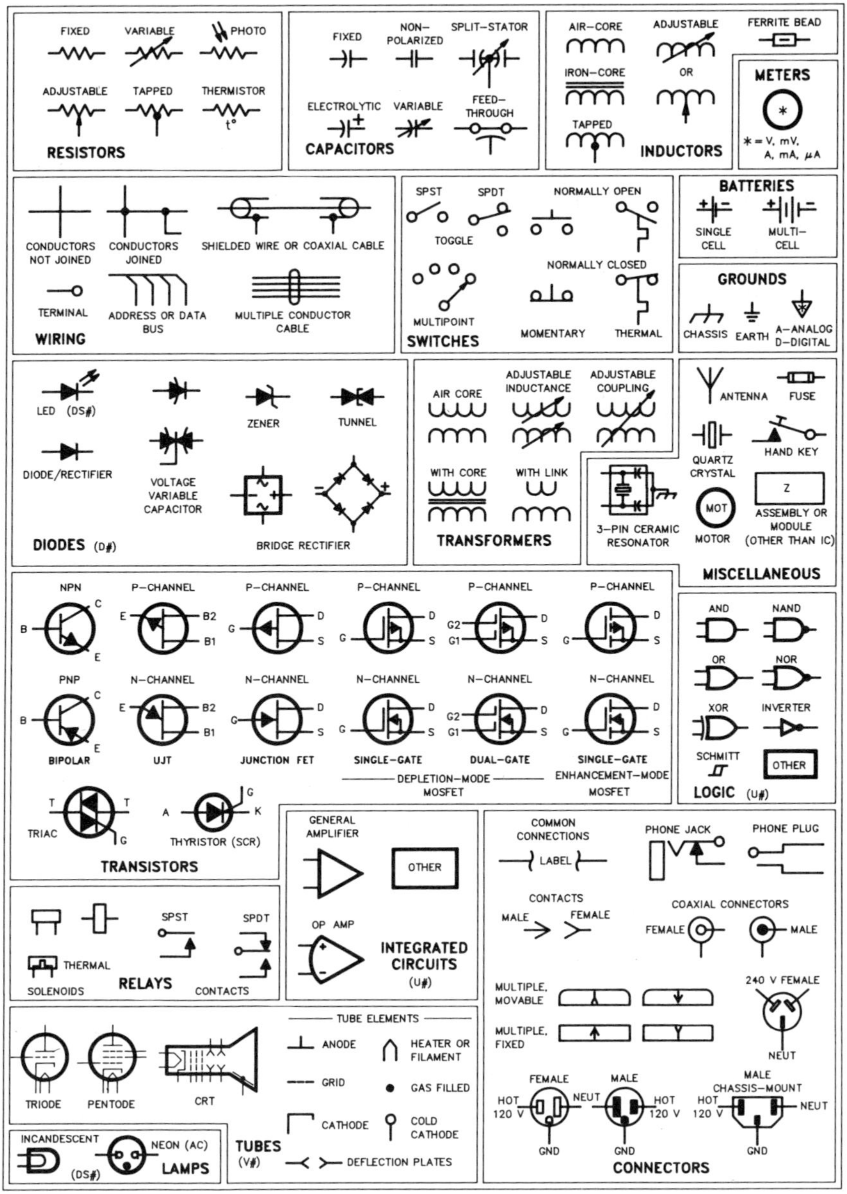

Standard Values for 1000-V Disc-Ceramic Capacitors

pF	*pF*	*pF*	*pF*
3.3	39	250	1000
5	47	270	1200
6	50	300	1500
6.8	51	330	1800
8	56	360	2000
10	68	390	2500
12	75	400	2700
15	82	470	3000
18	100	500	3300
20	120	510	3900
22	130	560	4700
24	150	600	5000
25	180	680	5600
27	200	750	6800
30	220	820	8200
33	240	910	10000

Common Values for Small Electrolytic Capacitors

μF	*V**	*μF*	*V**
33	6.3	10	35
33	10	22	35
100	10	33	35
220	10	47	35
330	10	100	35
470	10	220	35
10	16	330	35
22	16	470	35
33	16	1000	35
47	16	1	50
100	16	2.2	50
220	16	3.3	50
470	16	4.7	50
1000	16	10	50
2200	16	33	50
4.7	25	47	50
22	25	100	50
33	25	220	50
47	25	330	50
100	25	470	50
220	25	10	63
330	25	22	63
470	25	47	63
1000	25	1	100
2200	25	10	100
4.7	35	33	100

*Working voltage

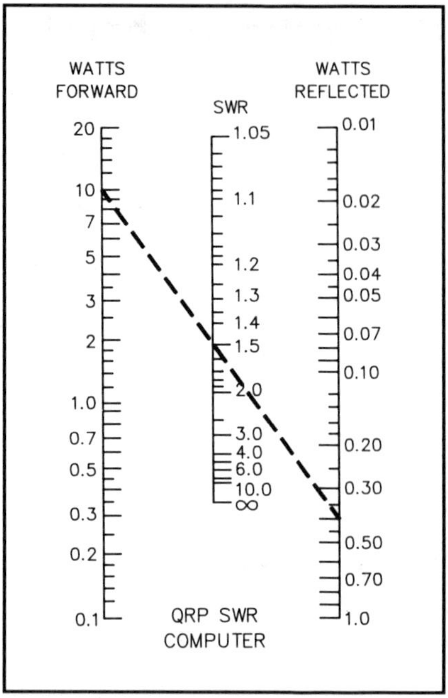

Nomograph of SWR versus forward and reflected power for levels up to 20 watts. Dashed line shows an SWR of 1.5:1 for 10 W forward and 0.4 W reflected.

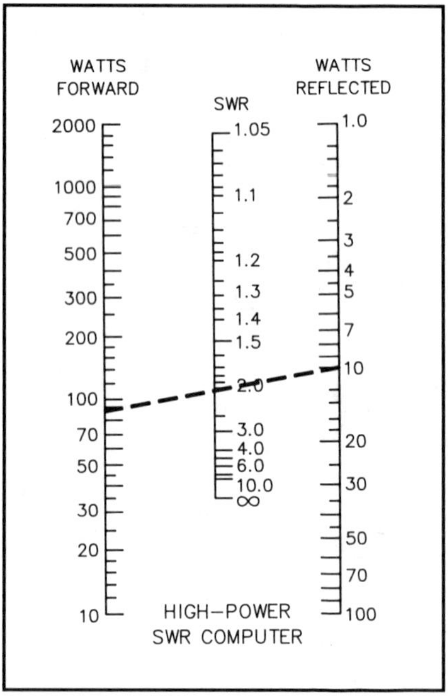

Nomograph of SWR versus forward and reflected power for levels up to 2000 watts. Dashed line shows an SWR of 2:1 for 90 W forward and 10 W reflected.

Fractions of an Inch with Metric Equivalents

Fractions Of An Inch		*Decimals Of An Inch*	*Millimeters*	*Fractions Of An Inch*		*Decimals Of An Inch*	*Millimeters*
	1/64	0.0156	0.397		33/64	0.5156	13.097
1/32		0.0313	0.794	17/32		0.5313	13.494
	3/64	0.0469	1.191		35/64	0.5469	13.891
1/16		0.0625	1.588	9/16		0.5625	14.288
	5/64	0.0781	1.984		37/64	0.5781	14.684
3/32		0.0938	2.381	19/32		0.5938	15.081
	7/64	0.1094	2.778		39/64	0.6094	15.478
1/8		0.1250	3.175	5/8		0.6250	15.875
	9/64	0.1406	3.572		41/64	0.6406	16.272
5/32		0.1563	3.969	21/32		0.6563	16.669
	11/64	0.1719	4.366		43/64	0.6719	17.066
3/16		0.1875	4.763	11/16		0.6875	17.463
	13/64	0.2031	5.159		45/64	0.7031	17.859
7/32		0.2188	5.556	23/32		0.7188	18.256
	15/64	0.2344	5.953		47/64	0.7344	18.653
1/4		0.2500	6.350	3/4		0.7500	19.050
	17/64	0.2656	6.747		49/64	0.7656	19.447
9/32		0.2813	7.144	25/32		0.7813	19.844
	19/64	0.2969	7.541		51/64	0.7969	20.241
5/16		0.3125	7.938	13/16		0.8125	20.638
	21/64	0.3281	8.334		53/64	0.8281	21.034
11/32		0.3438	8.731	27/32		0.8438	21.431
	23/64	0.3594	9.128		55/64	0.8594	21.828
3/8		0.3750	9.525	7/8		0.8750	22.225
	25/64	0.3906	9.922		57/64	0.8906	22.622
13/32		0.4063	10.319	29/32		0.9063	23.019
	27/64	0.4219	10.716		59/64	0.9219	23.416
7/16		0.4375	11.113	15/16		0.9375	23.813
	29/64	0.4531	11.509		61/64	0.9531	24.209
15/32		0.4688	11.906	31/32		0.9688	24.606
	31/64	0.4844	12.303		63/64	0.9844	25.003
1/2		0.5000	12.700	1		1.0000	25.400

B Equations Used In This Book

True forward power = Forward power reading – Reflected power reading (Equation 4-1)

$$R_{TOTAL} = R_1 + R_2 + R_3 + \cdots + R_n$$ (Equation 5-1)

$$R_{TOTAL} = \frac{1}{\frac{1}{R_1} + \frac{1}{R_2} + \frac{1}{R_3} + \cdots + \frac{1}{R_n}}$$ (Equation 5-2)

$$R_{TOTAL} = \frac{R_1 \times R_2}{R_1 + R_2}$$ (Equation 5-3)

$$\text{current} = \frac{\text{voltage}}{\text{resistance}}$$ (Equation 5-4)

$$\text{resistance} = \frac{\text{voltage}}{\text{current}}$$ (Equation 5-5)

$$\text{voltage} = \text{current} \times \text{resistance}$$ (Equation 5-6)

$$E = IR \ (\text{amperes} = \text{volts} \times \text{ohms})$$ (Equation 5-7)

$$I = \frac{E}{R} \ (\text{amperes} = \text{volts divided by ohms})$$ (Equation 5-8)

$$R = \frac{E}{I} \ (\text{ohms} = \text{volts divided by amperes})$$ (Equation 5-9)

$$L_{TOTAL} = L_1 + L_2 + L_3 + \cdots + L_n$$ (Equation 5-10)

$$L_{TOTAL} = \frac{1}{\frac{1}{L_1} + \frac{1}{L_2} + \frac{1}{L_3} + \cdots + \frac{1}{L_n}}$$ (Equation 5-11)

$$L_{TOTAL} = \frac{L_1 \times L_2}{L_1 + L_2}$$ (Equation 5-12)

$$C_{TOTAL} = C_1 + C_2 + C_3 + \cdots + C_n$$ (Equation 5-13)

$$C_{TOTAL} = \frac{1}{\frac{1}{C_1} + \frac{1}{C_2} + \frac{1}{C_3} + \cdots + \frac{1}{C_n}}$$ (Equation 5-14)

$$C_{TOTAL} = \frac{C_1 \times C_2}{C_1 + C_2}$$ (Equation 5-15)

$$Bw = 2 \times (D + M)$$ (Equation 8-1)

C Glossary of Key Words

Ammeter—A test instrument that measures current.

Amplifier—A device usually employing electron tubes or transistors to increase the voltage, current or power of a signal. The amplifying device may use a small signal to control voltage and/or current from an external supply. A larger replica of the small input signal appears at the device output.

ANSI RF protection guide—Standards formulated by the American National Standards Institute that provide RF exposure guidelines.

Antenna—A device that picks up or sends out radio waves.

Balanced antenna—An antenna that is symmetrical about the feed point. There are two equal sides to the antenna. A dipole is a balanced antenna.

Balun—Contraction for *bal*anced to *un*balanced. A device to couple a balanced load to an unbalanced source, or vice versa.

Band-pass filter—A circuit that allows signals to go through it only if they are within a certain range of frequencies. It attenuates signals above and below this range.

Bandwidth—The amount of space (measured in hertz) that a signal occupies.

Beam antenna—A directional antenna. A beam antenna must be rotated to provide coverage in different directions.

Beat-frequency oscillator (BFO)—An oscillator that provides a signal to the product detector. In the product detector, the BFO signal and the IF signal are mixed to produce an audio signal.

Break—An interruption of a QSO. When said twice over a repeater, it means "I must make a distress call immediately!"

Capacitance—The ability of a capacitor to store energy in an electric field.

Capacitor—An electrical component usually formed by separating two conductive plates with an insulating material. A capacitor stores energy in an electric field.

Closed—A repeater that restricts access to those who know a special code.

Coaxial cable—Coax (pronounced kó-aks). A type of feed line with one conductor inside the other.

Color code—A system in which numerical values are assigned to various colors. Colored stripes are painted on the body of resistors and sometimes other components to show their value.

Core—The material used in the center of an inductor coil.

Courtesy tone—A sound heard on a repeater indicating that a transmission is complete.

Critical frequency—The highest frequency at which a vertically incident radio wave will return from the ionosphere. Above the critical frequency, radio signals pass through the ionosphere instead of returning to Earth.

Crystal calibrator—(see **marker generator**).

Cubical quad antenna—An antenna built with its elements in the shape of four-sided loops.

CW—A communications mode transmitted by on/off keying of a radio-frequency signal. Another name for international Morse code.

D region—The lowest region of the ionosphere. Acts mainly to absorb energy from radio waves as they pass through it. This ionospheric absorption has a significant effect on signals below about 7.5 MHz during daylight.

Data—Computer-based modes, such as RTTY and packet radio.

Delta loop antenna—A variation of the cubical quad with triangular elements.

Detector—The stage in a receiver in which the modulation (voice or other information) is recovered from the RF signal.

Direct wave—A radio wave that travels in a straight line.

Directional wattmeter—(see **wattmeter**)

Director—An element in front of the driven element in a Yagi and some other directional antennas.

Driven element—The part of an antenna that connects directly to the feed line.

Dummy antenna (dummy load)—A station accessory that allows you to test or adjust transmitting equipment without sending a signal out over the air.

Duplex operation—Receiving and transmitting on two different frequencies. Also see **Simplex operation.**

E region—The second lowest ionospheric region. Exists only during the day. Under certain conditions, may refract radio waves enough to return them to Earth.

Electric field—An invisible force of nature. An electric field exists in a region of space if an electrically charged object placed in the region is subjected to an electrical force.

Electrolytic capacitor—A polarized capacitor that relies on a chemical reaction to form a thin insulating layer between the plates.

Emergency traffic—Messages concerning the immediate safety of human life or the immediate protection of property.

Emergency—A situation where there is a danger to lives or property.

Emission types—Term for the different modes authorized for use on the Amateur Radio bands. Examples are CW, SSB and FM.

F region—A combination of the two highest ionospheric regions, the F1 and F2 regions. Refracts radio waves and returns them to Earth. Its height varies greatly depending on the time of day, season of the year and amount of sunspot activity.

Farad—The basic unit of capacitance.

Feed line (feeder)—The wires or cable used to connect a transmitter or receiver to an antenna. (Also called *transmission line*.)

Film resistor—A resistor made by depositing a thin layer of resistive material on a ceramic form.

Filter—A circuit that will allow some signals to pass through it but will greatly reduce the strength of others.

Forward power—The power of the radio signal moving from the transmitter to the antenna along a **feed line**. (Also see **reflected power**.)

Frequency coordinator—An individual or group that recommends repeater frequencies.

Frequency deviation—The instantaneous change in frequency for a given FM or PM signal. Too much deviation (overdeviation or overmodulation) causes splatter.

Frequency discriminator—A circuit that produces a waveform which varies in amplitude as the incoming signal frequency changes. A discriminator is the **detector** in an FM receiver.

Frequency-modulated (FM) voice—Mode of voice (phone) communications used on repeaters.

Frequency modulation (FM)—The process of varying the frequency of an RF carrier in response to the instantaneous changes in the modulating signal.

Frequency separation—The difference between a repeater's receive and transmit frequencies.

Frequency-shift keyed telegraphy (FSK)—A method of transmitting radioteletype information by switching an RF carrier between two separate frequencies. FSK RTTY is most often used on HF. Intermediate-frequency (IF) filter—A circuit that limits the bandwidth of signals fed through the intermediate-frequency (IF) stage of a receiver.

Groundplane antenna—A vertical antenna mounted above ground, and with horizontal radial wires or a metal surface to form a simulated ground under the base of the antenna.

Health-and-Welfare traffic—Messages concerning a person's well being.

Henry—The basic unit of inductance.

High-pass filter—A filter that allows signals above the cutoff frequency to pass through. It attenuates signals below the cutoff frequency.

Horizontally polarized wave—An electromagnetic wave with its electric lines of force parallel to the ground.

Induced EMF—A voltage produced by a change in magnetic lines of force around a conductor. When a magnetic field is formed by current in the conductor, the induced voltage always opposes change in that current.

Inductance—The ability of a coil to store energy in a magnetic field.

Inductor—An electrical component usually composed of a coil of wire wound on a central core. An inductor stores energy in a magnetic field.

Input frequency—A repeater's receiving frequency. To use a repeater, transmit on the input frequency and receive on the output frequency.

Intermediate frequency (IF)—The output frequency of a mixing stage in a superheterodyne receiver. The subsequent stages in the receiver are tuned for maximum efficiency at the IF.

Ionosphere—A region of charged particles about 30 to 260 miles above the Earth that bends radio waves and returns them to Earth.

Ionospheric absorption—The D region of the ionosphere absorbs energy from radio waves that pass through it during the daytime. Limits communications range on the longer-wavelength amateur bands during the day.

Limiter—An FM-receiver stage between the IF amplifier and the **frequency discriminator**, which chops noise and amplitude variations from an incoming signal. The limiter produces a constant-amplitude signal for the discriminator.

Low-pass filter—A filter that allows signals below the cutoff frequency to pass through and attenuates signals above the cutoff frequency.

Marker generator—A high-stability oscillator that produces reference signals at known frequency intervals. It can be used to calibrate receiver and transmitter tuning dials. Also called a **crystal calibrator**.

Maximum usable frequency (MUF)—The highest-frequency radio signal that will reach a particular destination using E- or F-skip propagation. The MUF may vary for radio signals sent to different destinations.

MAYDAY—The distress call used on phone.

Modulate—To vary the amplitude, frequency, or phase of a radio-frequency signal.

Modulation—The process of varying a radio wave in some way to send information.

Morse code—(See CW.)

Multimeter—An electronic test instrument used to measure current, voltage and resistance in a circuit. Describes all meters capable of making these measurements, such as the *volt-ohm-milliammeter (VOM)*, *vacuum-tube voltmeter (VTVM)* and *field-effect transistor VOM (FET VOM)*.

National Electrical Code—A set of guidelines governing electrical safety, including antennas.

Nondirectional antenna—An antenna that sends radio signals equally well in all compass directions. A single vertical antenna, such as a **groundplane antenna** is an example of a nondirectional or *omnidirectional* antenna.

Ohm—The basic unit of electrical resistance, used to describe the amount of opposition to current.

Ohm's Law—A basic law of electronics. Ohm's Law gives a relationship between voltage (E), current (I) and resistance (R). The voltage applied to a circuit is equal to the current through the circuit times the resistance of the circuit (E = IR).

Open—A repeater that can be used by all properly licensed hams.

Open-wire feed line—Parallel-conductor feeder with air as its primary insulation material.

Output frequency—A repeater's transmitting frequency. To use a repeater, transmit on the **input frequency** and receive on the output frequency.

Packet radio—A system of digital communication whereby information is broken into short bursts. The bursts ("packets") also contain addressing and error-detection information.

Parallel circuit—An electrical circuit in which the electrons follow more than one path in going from the negative supply terminal to the positive terminal.

Parallel-conductor feed line—Feed line with two conductors held a constant distance apart.

Parasitic element—Part of a directive antenna that derives energy from mutual coupling with the driven element. Parasitic elements are not directly connected to the feed line.

Phase modulation (PM)—Varying the phase of an RF carrier in response to the instantaneous changes in the modulating signal.

Phone—Another name for voice communications, such as SSB and FM.

Polarization—Describes the electrical-field characteristic of a radio wave. An antenna that is parallel to the surface of the earth, such as a dipole, produces horizontally polarized waves. One that is perpendicular to the earth's surface, such as a quarter-wave vertical, produces vertically polarized waves.

Potentiometer—A resistor whose resistance can be varied continuously over a range of values.

Q signal—Three-letter symbol beginning with Q. Used on CW to save time and to improve communication. Some examples are QRS (send slower), QTH (location), QSO (ham conversation) and QSL (acknowledgment of receipt).

QRL?—Ham radio Q signal meaning "Is this frequency in use?"

QRZ?—Ham radio Q signal meaning "Is someone calling me?"

QSO—A conversation between two radio amateurs.

QSY—Ham radio Q signal meaning "change frequency."

Quieting—On repeaters, signal reports are given in terms of the amount of quieting. *Full quieting* means the signal is strong enough to overcome all receiver noise.

RACES (Radio Amateur Civil Emergency Service)—A part of the amateur service that provides radio communications only for civil defense purposes.

Radio-frequency (RF) carrier—A constant-amplitude, unmodulated, radio-frequency signal.

Radio Frequency Protection Guide of the American National Standards Institute—Standards that recommend RF exposure limits for humans.

Radioteletype (RTTY)—Radio signals sent from one teleprinter machine to another machine. Anything that one operator types on his teleprinter will be printed on the other machine. Also known as narrow-band direct-printing telegraphy.

Reactance—The property of an inductor or capacitor (measured in ohms) that impedes current in an ac circuit without converting power to heat.

Reactance modulator—A device capable of modulating an ac signal by varying the reactance of a circuit in response to the modulating signal. (The modulating signal may be voice, data, video, or some other kind depending on what type of information is being transmitted.)

Reflected power—The power of the radio signal moving from the antenna to the transmitter along a **feed line**. The power of the signal *reflected* from the antenna because of an impedance mismatch. (Also see **forward power**.)

Reflectometer—A test instrument used to indicate standing wave ratio (SWR) by measuring the forward power (power from the transmitter) and reflected power (power returned from the antenna system).

Reflector—An element behind the driven element in a Yagi and some other directional antennas.

Repeater—An amateur station that receives a signal and retransmits it for greater range.

Resistance—The ability to oppose an electric current.

Resistor—Any material that opposes a current in an electrical circuit. An electronic component specifically designed to oppose current.

RF burn—A flesh burn caused by exposure to a strong field of radio-frequency (RF) energy.

RST—A system of numbers used for signal reports: R is readability, S is strength and T is tone. (On single-sideband phone, only R and S reports are used.)

S meter—A meter in a receiver that shows the relative strength of a received signal.

Scatter—A small amount of signal that is reflected from the Earth's surface after traveling through the ionosphere. The reflected signals may go back into the ionosphere along several paths and be refracted to Earth again. Scatter can help provide communications into a station's skip zone.

Series circuit—An electrical circuit in which all the electrons must flow through every part of the circuit. There is only one path for the electrons to follow.

Sidebands—The sum or difference frequencies generated when an RF carrier is mixed with an audio signal. Single-sideband phone (SSB) signals have an upper sideband (USB—that part of the signal above the carrier) and a lower sideband (LSB—the part of the signal below the carrier). SSB transceivers allow operation on either USB or LSB.

Signal generator—A test instrument that produces a stable low-level radio-frequency signal. The signal can be set to a specific frequency and used to troubleshoot RF equipment.

Signal report—An exchange between stations indicating the relative strength of each other's signals.

Simplex operation—Receiving and transmitting on the same frequency. See **Duplex operation.**

Single sideband (SSB)—Emission mode that describes the type of voice emission used most on the HF bands.

Skip zone—An area past the maximum range of ground waves and before the range of waves returned from the ionosphere. Radio communication is difficult or impossible between stations in each other's skip zone.

$\overline{\text{SOS}}$—The distress call used on CW.

Splatter—A type of interference to nearby stations. Occurs when a transmitter is overmodulated.

Standing-wave ratio (SWR)—Sometimes called VSWR. A measure of the impedance match between the feed line and the antenna. Also, with a Transmatch in use, a measure of the match between the feed line from the transmitter and the antenna system. The system includes the Transmatch and the line to the antenna. VSWR is the ratio of maximum voltage to minimum voltage along the feed line. Also the ratio of antenna impedance to feed-line impedance when the antenna is a purely resistive load.

SWR meter—(see **reflectometer**)

Tactical call signs—Type of station identification used to help coordinate public-service events. Examples are *parade headquarters* and *net control*.

Temperature inversion—A condition in the atmosphere in which a region of cool air is trapped beneath warmer air.

Terminal node controller (TNC)—A device that accepts information from a computer and converts the information into packets. The TNC also receives packets and extracts information to be displayed by a computer.

Test emission—Transmissions containing no information, such as an unmodulated carrier.

Time-out timer—A device that limits the amount of time any one person can talk through a repeater.

Tone-modulated CW (MCW)—A signal produced by feeding an audio tone into the microphone jack of an FM transmitter, and then keying the tone on and off to send Morse code.

Toroidal core—A donut-shaped form used to hold the turns of an inductor. Toroidal cores are usually made from ferrite or powdered-iron materials.

Troposphere—The region in Earth's atmosphere just above the Earth's surface and below the ionosphere.

Tropospheric bending—When radio waves are bent in the troposphere, they return to Earth farther away than the visible horizon.

Tropospheric ducting—A type of VHF propagation that can occur when there is a temperature inversion.

Unbalanced line—Feed line with one conductor at ground potential, such as coaxial cable.

Variable capacitor—A capacitor that can have its value changed within a certain range.

Variable-frequency oscillator (VFO)—An oscillator used in receivers and transmitters. The frequency is set by a tuned circuit using capacitors and inductors. The frequency can be changed by adjusting the components in the tuned circuit.

Variable resistor—A resistor whose value can be adjusted over a certain range.

Vertically polarized wave—A radio wave that has its electric lines of force perpendicular to the surface of the earth.

Visible horizon—The most distant point one can see by line of sight.

Voice communications—Hams can use several voice modes, including FM and SSB.

Voltmeter—A test instrument used to measure voltage.

Wattmeter—Also called a power meter, a test instrument used to measure the power output (in watts) of a transmitter. A directional wattmeter measures both forward and reflected power.

Wire-wound resistor—A resistor made by winding a length of wire on an insulating form.

WWV and **WWVH**—Radio stations that broadcast the precise time on precise frequencies. Hams tune in to these stations to calibrate transceiver dials (and to set their clocks).

Yagi antenna—The most popular type of amateur directional (beam) antenna. It has one driven element and one or more additional elements.

About the
American Radio Relay League

The seed for Amateur Radio was planted in the 1890s, when Guglielmo Marconi began his experiments in wireless telegraphy. Soon he was joined by dozens, then hundreds, of others who were enthusiastic about sending and receiving messages through the air—some with a commercial interest, but others solely out of a love for this new communications medium. The United States government began licensing Amateur Radio operators in 1912.

By 1914, there were thousands of Amateur Radio operators—hams—in the United States. Hiram Percy Maxim, a leading Hartford, Connecticut, inventor and industrialist saw the need for an organization to band together this fledgling group of radio experimenters. In May 1914 he founded the American Radio Relay League (ARRL) to meet that need.

Today ARRL, with about 160,000 members, is the largest organization of radio amateurs in the United States. The League is a not-for-profit organization that:

- promotes interest in Amateur Radio communications and experimentation
- represents US radio amateurs in legislative matters, and
- maintains fraternalism and a high standard of conduct among Amateur Radio operators.

At League headquarters in the Hartford suburb of Newington, the staff helps serve the needs of members. ARRL is also International Secretariat for the International Amateur Radio Union, which is made up of similar societies in more than 100 countries around the world.

ARRL publishes the monthly journal *QST*, as well as newsletters and many publications covering all aspects of Amateur Radio. Its headquarters station, W1AW, transmits bulletins of interest to radio amateurs and Morse-code practice sessions. The League also coordinates an extensive field organization, which includes volunteers who provide technical information for radio amateurs and public-service activities. ARRL also represents US amateurs with the Federal Communications Commission and other government agencies in the US and abroad.

Membership in ARRL means much more than receiving *QST* each month. In addition to the services already described, ARRL offers membership services on a personal level, such as the ARRL Volunteer Examiner Coordinator Program and a QSL bureau.

Full ARRL membership (available only to licensed radio amateurs) gives you a voice in how the affairs of the organization are governed. League policy is set by a Board of Directors (one from each of 15 Divisions). Each year, half of the ARRL Board of Directors stands for election by the full members they represent. The day-to-day operation of ARRL HQ is managed by an Executive Vice President and a Chief Financial Officer.

No matter what aspect of Amateur Radio attracts you, ARRL membership is relevant and important. There would be no Amateur Radio as we know it today were it not for the ARRL. We would be happy to welcome you as a member! (An Amateur Radio license is not required for Associate Membership.) For more information about ARRL and answers to any questions you may have about Amateur Radio, write or call:

ARRL Educational Activities Dept
225 Main Street
Newington CT 06111-1494
(203) 666-1541
Prospective new amateurs call:
800-32-NEW HAM (800-326-3942)

Index

A

B

C

D

E

F

G

H

I

L

M

N

O

P

Q

R

S

T

U

V

W

Y

Numbers

Notes

TECHNICIAN
CLASS LICENSE
MANUAL

PROOF OF
PURCHASE

FEEDBACK

Please use this form to give us your comments on this book and what you'd like to see in future editions.

License class:

□ Novice □ Technician □ Technician with HF privileges

□ General □ Advanced □ Extra

Name ______________________ ARRL member? □ Yes □ No

Call sign ______________________

Daytime Phone ( ) ______________________ Age ______________________

Address ______________________

City, State/Province, ZIP/Postal Code ______________________

If licensed, how long? ______________________

Other hobbies ______________________

____

Occupation ______________________

For ARRL use only

Edition 2 3 4 5 6 7 8 9 10 11 12

Printing 1 2 3 4 5 6 7 8 9 10 11 12

From ______________________

____

____

Please affix postage. Post Office will not deliver without postage.

EDITOR, TECHNICIAN CLASS LICENSE MANUAL
AMERICAN RADIO RELAY LEAGUE
225 MAIN ST
NEWINGTON CT 06111

.... please fold and tape ....